D0207269

Sustainable Strategic Management

W. Edward Stead
and Jean Garner Stead
with Mark Starik

M.E.Sharpe
Armonk, New York
London, England

Copyright © 2004 by M.E. Sharpe, Inc.

All rights reserved. No part of this book may be reproduced in any form
without written permission from the publisher, M.E. Sharpe, Inc.,
80 Business Park Drive, Armonk, New York 10504.

Library of Congress Cataloging-in-Publication Data

Stead, W. Edward
Sustainable strategic management / by W. Edward Stead and Jean Garner Stead,
with Mark Starik.
p. cm.
Includes bibliographical references and index.
ISBN 0-7656-1131-7 (cloth: alk. paper) ISBN 0-7656-1132-5 (pbk: alk. paper)
1. Sustainable development. 2. Strategic planning. 3. Management.
I. Stead, Jean Garner. II. Starik, Mark. III. Title.
HC79.E5 S717 2004
658.4′012—dc21

2003012436

Printed in the United States of America

The paper used in this publication meets the minimum requirements of
American National Standard for Information Sciences
Permanence of Paper for Printed Library Materials,
ANSI Z 39.48-1984.

BM (c) 10 9 8 7 6 5 4 3 2 1
BM (p) 10 9 8 7 6 5 4 3 2 1

In memory of our dear friend Rick Watson (1943–2002)
We miss you so much.

Dedicated to our friends and colleagues in the
Organizations and the Natural Environment (ONE) Interest Group
and the Social Issues in Management (SIM) Division
of the Academy of Management

Contents

List of Tables and Figures

Tables

Figures

List of Tables and Figures

Tables

Figures

Preface

Since the field of organizations and the natural environment emerged in the early 1990s, tremendous progress has been made in the search for sustainable strategic management systems in business organizations. On the practice side, organizations, which have historically been major contributors to environmental and social problems, now seem primed to become major parts of finding solutions to these problems. Environmentally, organizations today regularly apply environmental management techniques such as eco-efficiency, pollution prevention, total quality environmental management, and design for environment, and many are making concerted efforts to meet and exceed external environmental standards, such as the World Business Council on Sustainable Development's ecoefficiency criteria, the ISO 14000 process, and the CERES Principles. Socially, organizations in today's business world have embraced the concept of stakeholder management, and many have incorporated corporate social responsibility into the core of their strategies, reporting their social performance to stakeholders via corporate citizenship reports and so forth. External standards such as those set by Business for Social Responsibility, the United Nations Declaration of Human Rights, and the International Labour Organization are making being a good corporate citizen a basic business requirement.

On the academic side, environmental management has matured into a legitimate research field and curriculum path in business schools across the globe. The explosion of literature related to improved environmental performance and the emergence of organizations such as the Academy of Management's Organizations and the Natural Environment Interest Group and the Greening of Industry Network (GIN) have been instrumental in advancing both legitimacy and knowledge in the field. Of course, concern for the social performance of business has been around for some time. The Social Issues in Management (SIM) Division of the Academy of Management was one of the Academy's original divisions, and since its inception over thirty years ago, it has fostered strong research agendas related to corporate social responsibility, social performance, ethics, public policy, stakeholder theory, the environment, and so forth. Since SIM was founded, several other organizations that focus on social issues in business have emerged, including the International Association of Business and Society and the Society for Business Ethics.

The concept of sustainability provides the framework to integrate both the environmental and social performance of businesses under the same umbrella. In practice, in research, and in the classroom, it makes sense to incorporate economic, environmental, and social performance into the organization's strategic management systems and processes. With this in mind, we attempt in this book to develop a comprehensive theory of sustainable strategic management that expands the traditional strategic management process to include both nature and society in the formulation, implementation, and evaluation of strategies in business organizations. It is our hope that this framework will provide students, practitioners, environmental professionals, and academics with ways to integrate a core value of sustainability into their business operations, studies, academic research agendas, and everyday lives.

Many people have contributed in so many different ways to the completion of this book. First of all, we would like to thank Mark Starik for his contributions. We would also like to thank Phil Miller, chairperson of East Tennessee State University's Department of Management and Marketing, and Linda Garceau, dean of East Tennessee State University's College of Business. Both have provided the academic environment and support we have needed to pursue our research and teaching agendas. We especially appreciate their support of Jean Stead's noninstructional leave in the spring of 2003. Without that time, completing this book would have been very difficult.

We would like to thank our graduate assistants, Mark Lundy, Amanda Smith, and Melissa Smith, for their high quality assistance. We would also like to thank Cynthia Taylor for her help in gathering and organizing the data on Appalachian Sustainable Development and Ford Motor Company, helping to write the introductory narratives in Chapter 1, and editing the early chapters of this book. Thanks are certainly in order to Craig Turner, who took time to read this manuscript and give us his thoughtful feedback. And we give special thanks to Anthony Flaccavento, the Board of Directors, and the staff of Appalachian Sustainable Development (ASD) for their approval, time, and help in preparing the ASD case that serves to open and close this book.

We also want to extend our most heartfelt gratitude to our friends and colleagues in the Organizations and the Natural Environment Interest Group and the SIM Division of the Academy of Management. They have been the core of our professional family for almost a quarter of a century. Our associations with these people have not only contributed tremendously to our professional development; they have also enriched our lives.

Jean would like to express her most sincere appreciation to the "Paradise Isle Gang" in Riverside, Alabama (Helen White, Jenny and Bud Moore, Joel and Leesa Durrett, Betty and Ed Galloway, Debbie and Steve Allen, Woody Barnes, Karen Clardy, Betty Minor, Pokey and Tommie Jean Forrester, the Judge, and many others), for providing many meals and laughs for her while she was on leave writing this book.

We want to give special thanks to our editor at M.E. Sharpe, Harry Briggs. This is the second book that we have written under Harry's watchful eye, and we can attest that on both occasions he was encouraging, supportive, professional, and helpful. The work of good editors goes unnoticed most of the time because it is all done behind the scenes, but we can assure you that editors have a huge impact on the final products generated by their authors. Thanks, Harry, for having faith in us and for supporting us in our efforts. In our opinion, you are the best editor on the small planet earth.

Finally, we would like to thank our daughter, Garner Lee Stead, for being the person she is. She was only ten years old when we began writing our first book, and she has been suffering the slings and arrows of our book writing efforts ever since. Now twenty-three years old, she is in the process of establishing her own life in Indianapolis, hundreds of miles away from Mom and Dad. But even from that distance, she is always there for us. Thanks, Garner Lee, for your support and your unconditional love. When he looked into your eyes when you were born, Ed said that you were a young person with an old soul. We only hope that someday we can be as wise as you.

Peaceful solutions to all,
Ed and Jean

Part I

Foundations of Sustainable Strategic Management

Our purpose in this section is to set the stage for sustainable strategic management. We begin by introducing two very committed sustainable strategic managers, Anthony Flaccavento of Appalachian Sustainable Development and Bill Ford of Ford Motor Company. We use their stories to illustrate that sustainable strategic management is not just a new way of doing business; it is a new way of thinking about business, and adopting it will require widespread transformational change in business organizations. We then define sustainability, the key concept underlying this book, focusing on its scientific principles, its key dimensions, and its ethical foundations. We conclude this section by defining sustainable strategic management, relating it to the popular triple-bottom-line concept, and discussing the roles and responsibilities of CEOs and boards of directors in leading organizations toward sustainable strategic management.

1

The Winds of Change

Anthony Flaccavento was born in New York in 1957 and raised in Baltimore by politically liberal, middle-class parents who instilled in him a love of music, a respect for people over things, a sense of frugality, and a love for the land. His parents were early recyclers, and his father always had an organic garden. When Flaccavento was ten years old, his father bequeathed him a four-by-four plot in his garden, and Flaccavento's lifetime passion for organic farming was born.

Flaccavento followed his passion to the University of Kentucky where he pursued a double major in agriculture and environmental science. He was heavily influenced during that time by the writings of Rachel Carson and Wendell Berry, and it did not take him long to discover the huge philosophical gulf between traditional agriculture, with its emphasis on super yields and chemical control, and environmental science, with its emphasis on ecological health and natural processes. In an effort to walk the line between his majors, he concentrated his studies in natural resource economics. After graduating, Flaccavento took a job in the rural coal mining counties in the Appalachian Mountains of Kentucky taking soil and water samples for an engineering company. This was his first exposure to the Appalachian region, a place he would eventually call home.

He left his job to pursue a master's in International Affairs with an emphasis on social and economic development at the University of Pittsburgh. After earning his degree, he went to work for the Catholic Social Justice Movement and was assigned to the Catholic Diocese in Richmond, Virginia, to manage the Appalachian Catholic Office of Justice and Peace in Castlewood, Virginia, a rural mountain community in the Clinch River Valley of southwest Virginia. During his tenure with the Office of Justice and Peace from 1985 to 1995, Flaccavento helped establish the First Homeowners program and the Share the Food program and helped found the Coalition for Jobs and the Environment (CJE).

Flaccavento served as the director of the CJE in its early years, and it was during that time that he struggled with the commonly held perception that jobs and the environment are always at odds. He saw unemployment rates as high as 17 percent in the region, and he saw a declining natural environment at the same time. Neither was benefiting from the jobs-versus-nature mentality. He strongly believed that the ultimate purpose of an economy is to take care of its people, and this belief along with his burning desire to pursue sustainable agriculture led him to rethink his ideas about sustainability. He and some colleagues began a yearlong community strategic planning process that included developing principles, benchmarks, and evaluative tools for sustainable management of agriculture, and the seeds of Appalachian Sustainable Development (ASD), his present enterprise, were planted.

ASD, a not-for-profit organization, was founded in 1995 as a result of that yearlong planning process. The organization works in ten counties in the Appalachian Mountains of Virginia and Tennessee. ASD focuses on developing healthy, diverse, and ecologically sound economic opportunities for local citizens through education, training, and the development of cooperative networks and marketing systems. ASD operates two strategic business units: Appalachian Harvest, which processes, packages, and distributes locally grown organic produce, and Sustainable Woods, which harvests, processes, and sells sustainable wood products.

Bill Ford was raised in different economic surroundings from Anthony Flaccavento's. Born William Clay Ford II in 1957, he was raised as an heir to the Ford Motor Company fortune, and, as it would turn out, heir to its management as well. Though Ford was born into a rich and powerful family, his parents, who were known for their kindness and humility, raised him as normally as possible, instilling in him basic egalitarian values. His great-grandfather, Henry Ford, was an avid bird-watcher, nature lover, and supporter of the National Park system, and Bill Ford developed his own love of nature during his annual trout fishing trips when he was a boy. He was also an athlete, and he chose to play hockey with working-class boys in another part of Detroit rather than with the wealthy children of his own neighborhood. A very good student, he believed that sports and academics were the two arenas where social standing counted the least and ability and desire the most.

Bill Ford attended Princeton as an undergraduate and received his master's in Management from the Massachusetts Institute of Technology (MIT). A theme that ran throughout his academic career was his struggle to understand how to simultaneously balance the interests of workers and trade unions with the interests of shareholders. He wrote an essay at Princeton questioning how his great-grandfather could pay the best wages in the world to his workers and yet be so antiunion, and he conversed regularly with MIT professor Tom Barconni about this issue while he was in graduate school. This struggle between the social interests of the workers and the economic interests of the firm seems to be a defining experience in the development of Bill Ford's management philosophy, the one he would later bring with him to Ford Motor Company.

Ford Motor Company chose Bill Ford to be its chief executive officer in October 2001, two years after he was appointed chairman of the board. Often referred to as a rebel, radical, Bolshevik, and idealist by detractors in the auto industry, Bill Ford has pledged the company's commitment to becoming a leader in corporate citizenship and making the automobile industry sustainable. He has a personal vision of achieving new levels of economic success for Ford Motor Company while at the same time establishing clear leadership in social and environmental issues. He has publicly committed the company to providing superior returns to stockholders and meeting customer demands in ways that do the least harm or most benefit to the environment and to society in general. He has initiated a process of stakeholder engagement by establishing dialogue with outside thinkers on social and environmental issues that can place Ford Motor Company in a leadership position in climate change, human rights, and corporate citizenship. By initiating the company's annual *Corporate Citizenship Report*, the first of which explicitly revealed the company's shortcomings in stewardship, he reinforced a commitment to stakeholders of corporate accountability and transparent operations.

In a speech delivered at the Fifth Annual Greenpeace Business Conference in October 2000, Bill Ford presented a plan that would establish Ford Motor Company as a leader in corporate citizenship, emphasizing that an automobile company must demonstrate leadership in addressing environmental concerns. He prefaced the plan with his belief that enlightened corporations "can only be as successful as the communities, and the world, that they exist in." Further, he is working to put his words into action at Ford Motor Company. His great-grandfather built the Rouge Complex in Dearborn, Michigan, the assembly line that revolutionized manufacturing and helped transform society in the twentieth century. Bill Ford is now leading the manufacturing world into a new manufacturing paradigm appropriate to the twenty-first century with the renovation of the Rouge. His goal is to transform the complex into a modern, integrated manufacturing system that is a world-class leader of lean and environmentally sensitive manufacturing. He said, "I'm in this for my children and my grandchildren. I want them to inherit a legacy they're proud of."[1]

Anthony Flaccavento and Bill Ford are certainly two very different strategic managers from very different backgrounds running very different organizations. Yet their stories are remarkably similar in so many ways. They both have dreams of achieving economic success for their organizations via processes that protect and enhance the planet's natural resources and the welfare of its citizens. Both are committed to leaving future generations a legacy to be cherished rather than condemned. Both view their role as one of change agent helping to shepherd the current generation into a new world that is able to integrate rather than segregate economic, social, and ecological health. Anthony Flaccavento and Bill Ford represent a new generation of strategic managers that not only does business in different ways, but also thinks and dreams in different ways. These gentlemen exemplify the underlying assumptions, values, and processes of *sustainable strategic management* (SSM), strategic management processes that are economically competitive, socially responsible, and in balance with the cycles of nature. Whereas in traditional strategic management, the term "sustainable" is typically used in reference to a firm's ability to continuously renew itself in order to survive over the long term, in this book we take a more comprehensive global view of the term, referring not only to the survival and renewal of the firm itself but also to the survival and renewal of the greater economic system, social system, and ecosystem in which the firm is embedded.

Will Bill Ford and Anthony Flaccavento ever realize their dreams? Will they be able to meet their economic goals in today's highly competitive and unforgiving business world by managing their organizations in socially and ecologically responsible ways, or will they have to compromise their dreams because their social and ecological concerns conflict with the economic realities of their marketplace? We have no definitive answers to these questions, but we do believe that the underlying concepts, frameworks, tools, and processes of sustainable strategic management provide pathways to the fulfillment of dreams of strategic managers like Bill Ford and Anthony Flaccavento. We wrote this book with the intent of revealing and examining these concepts, frameworks, tools, and processes in the hope that the book will serve to guide strategic management students, practitioners, and scholars as they search for ways to integrate economic, social, and ecological well-being for this generation and the generations to come.

The Earth Is a Small Planet

On the outer edges of a spinning cluster of stars and other heavenly bodies called the Milky Way lies a blue-green planet known as the earth. The earth has evolved from a hot, gaseous ball to a planet rich with the ingredients of life—breathable air, fresh water, abundant minerals, and so on—in approximately 4.5 billion years.

Single-celled life began on earth some 3.8 billion years ago, and today there are tens of millions of species living on the planet. The coevolutionary processes of the earth have resulted in an incredible ecological balance between plant life, animal life, and the planetary systems that support both. The earth is now dominated by a species known as Homo sapiens—human beings. The uniqueness of this species can be recognized in its name. The word sapiens comes from the Latin verb *sapere*, which means to be wise. Thus, this is the intelligent species, the one with the brains. The human brain is so powerful that it can design and build machines able to store and process millions of times the amount of information the brain itself can handle. Frederick Taylor, father of scientific management, saw perfect workers as extensions of their machines; on the contrary, humans' machines are extensions of themselves.

Before the beginning of the Industrial Revolution around 1650, people employed their machines in very limited ways, using wheels for wagons and grain mills, for example. Most of the population was engaged in agriculture, and most people spent their time simply trying to survive. Life was very hard and tenuous, and population growth was slow. Then came the Industrial Revolution with its powerful fossil fuel energy sources, mass production techniques, and modern transportation and communication systems. Over the past 350 years humankind moved away from older forms of society whose primary activities were hunting, raising livestock, planting, gathering, and milling to the modern industrial society of today. During that time, survival has become much easier and more secure for those in the world with a modicum of economic wealth. Modern farming techniques have made it possible to raise more food than humans can consume. High-speed communications and transportation have made it possible to speak with almost anyone in the world in a matter of seconds and carry on a face-to-face conversation with them in a matter of hours. Modern medicine has transformed deadly illnesses of the past century into minor irritations today. During this time the human species has survived and thrived on the planet, growing in numbers to 6.2 billion people.

However, the earth is not getting any bigger. This beautiful blue-green marble is still only 25,000 miles in circumference, 75 percent water, and much of the rest uninhabitable mountain, desert, and frozen tundra. The earth's natural resources are being depleted, and wastes are being generated at rates unheard-of in human history. Tropical forests are being cleared to make way for economic progress, water tables are being drawn down to dangerous levels throughout the world, and soil erosion is exceeding soil replenishment rates. Like the lemmings of arctic Norway, humans are breeding excessively and using their resources extravagantly.

All of this would be well and good if it were not for the fact that the life-giving and life-supporting processes of the earth are currently operating in a rather closed ecosystem. The more open a system is—that is, the more it can exchange energy, information, and wastes with its environment—the more renewing it can be. This is because the open system is able to import sufficient amounts of energy from its environment to replenish what it loses when it transforms its own energy, and it can expel the waste products that result from this transformation back into the environment. However, the more closed a system is, the less renewing it can be because it can neither import sufficient quantities of energy to replace its depleted resources, nor can it develop sufficient capacity to dispose of its wastes.

Open and closed are relative terms. As a system, the earth has only one significant energy input from its environment—solar flow, the sun. Through photosynthesis, solar energy provides the earth with the power to feed its species; it also provides the basic energy for water and wind cycles. The remainder of the planet's energy is tied directly to terrestrial resources and processes—oil, coal, wood, natural gas, uranium, and so forth. Further, the earth must absorb the wastes generated when energy is converted into something useful. These wastes are often buried in the ground, dumped in the water, or spewed into the air. For all but the past 350 years of the 4.5 billion-year history of the earth, these mechanisms provided a more than adequate amount of openness to meet the needs of life on the planet. It has only been during the Industrial Age that humankind has been using the earth's resources and discharging wastes at rates faster than renewal can take place. This means that the balance the planet once enjoyed has in 350 years—a split second of eternity—almost disappeared. Humankind has embarked on an experiment unprecedented in its long history, and, as it is beginning to discover, the results of this experiment could be disastrous.

Thus, the earth is a small planet, one that is responsible for supporting a rapidly growing human population seeking the improved lifestyles promised by the Industrial Revolution. By 2050 this planet may have 8 billion people or so, and that puts humankind in a somewhat mathematical bind; like the lemmings, humans are doing everything possible to squeeze more and more from less and less. In the process, the earth's natural resources are being expended faster than they can be renewed. The population of the planet has engaged willingly in an unprecedented economic experiment with an ecosystem that has evolved in a beautiful symphonic balance for 4.5 billion years. Although it has been known for years that the results for the human species may be nothing short of disastrous if civilization fails to adjust its rates of economic activity to the evolutionary processes of nature, the experiment continues at full throttle. Many believe that humans can save themselves from these problems with new technologies; however, this promise remains unfulfilled. Real change will require new values and new ways of thinking.

It's Time for a Change

Humankind has reached a point in its history where it needs to reassess where it is going and how it will get there. For the past 350 years, humans have built their hopes and dreams on the concept of unlimited economic growth. This unlimited growth paradigm has led to the belief that more production and consumption are good regardless of the environmental and social consequences. Gross domestic product (GDP) is viewed under this paradigm as a measure of pure good regardless of what actually constitutes it. Indeed, the desire for economic growth has been raised to mythic proportions that rival any religion in human history. In today's world, personal and societal welfare are measured almost solely on the amount of growth experienced in personal and societal wealth.

Yet, as humans continue at breakneck speed to produce and consume, produce and consume, produce and consume, they do so with the knowledge that they are threatening the very resources and systems that support quality human life on the planet. If humans continue to foul the air and water, degrade the land, ignore the poor and disenfranchised, exploit the natural beauty, and so on, they are in danger of leaving a world to their future generations that is not as hospitable to a quality human life as the one they inherited from their ancestors. Now is the time to break free from the mythic drive for economic growth. Humans are in need of a new economic paradigm based on the image of an interconnected world community of people functioning in harmony with one another and with nature.

The Closed Circular Flow Economy: Paradigm of the Past

At the heart of the paradigm that has guided strategic decision makers over the past 350 years is the image of the economy as a closed circular flow (see Figure 1.1). This framework depicts the production-consumption cycle in which resources are transformed by business organizations into products and services that are purchased by consumers. Note in the model that resources flow in one direction and money flows in the other.

The problem with the closed circular flow model is not what it depicts, but what it ignores. Depicting the economy as closed implicitly assumes that the economy is isolated and independent from the social system and ecosystem. Under such an assumption, the economy is not subject to the physical laws of the universe, the natural processes and cycles of the ecosystem, or the values and expectations of society. Assuming these away leaves humankind with a mental model of an economy that can grow forever as self-serving, insatiable consumers buy more and more stuff from farther and farther away

Figure 1.1 **The Closed Circular Flow Economy**

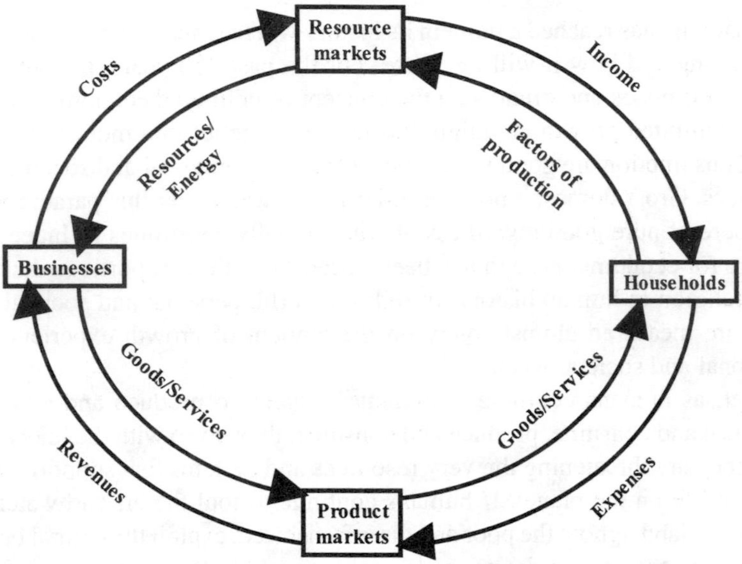

to satisfy a never-ending list of economic desires without any serious social and ecological consequences.

The Open Living System Economy: Paradigm of the Future

Unfortunately the economy is not isolated from the earth's other subsystems as depicted in the closed circular flow model. Long-term economic health can exist only within a social system and ecosystem that support it. Global economic activity must function within the natural and social boundaries of the planet. The earth is the ultimate source of natural and human capital for the economic system. Thus, a healthy flow of economic activity can be sustained for posterity only if strategic decision makers operate under the paradigm that the economy serves the needs of the greater society within the limits of nature.

Figure 1.2 depicts such a paradigm. In this model, the earth is a living system. As such, its survival is dependent on achieving a sustainable balance within its various subsystems. The most basic living subsystems on the planet are the individual organisms that inhabit it, and the most dominant of the planet's individual organisms is humankind. Because of this dominance, the decisions made by human beings are major forces influencing the ultimate state of society and nature. The ability of human beings to make effective decisions about how they interact with one another and with the planet depends on the accuracy of the mental processes they use to make those choices.

Figure 1.2 **The Open Living System Economy**

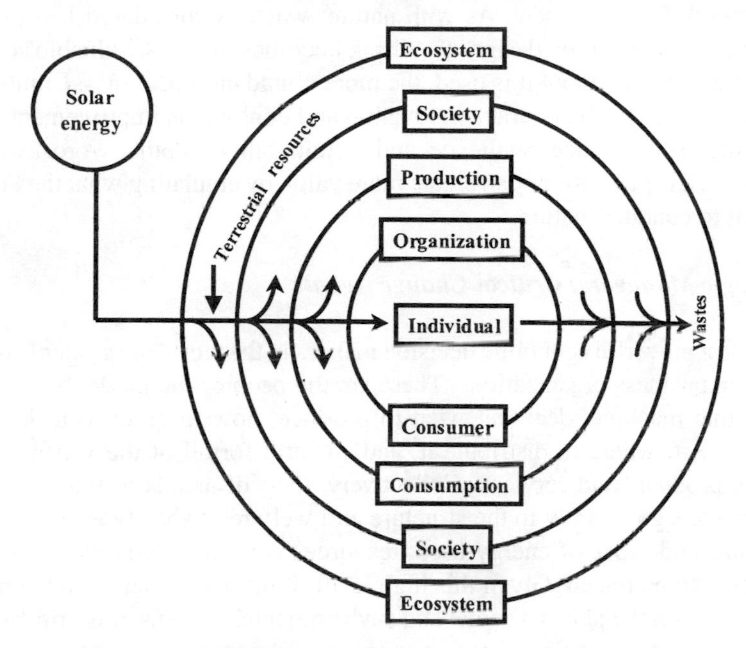

Source: W.E. Stead and J.G. Stead, *Management for a Small Planet* (Thousand Oaks, CA: Sage, 1996).

At the heart of these mental processes are the underlying assumptions and values human beings hold. Thus, human values and assumptions have a huge influence on the state of society and nature.

Of course, humans make their decisions in a variety of collective contexts. Decisions are made in the context of families, business organizations, educational institutions, governmental agencies, and interest groups, to name but a few. In the economic realm, individuals make decisions as members of organizations that produce goods and services, and they make decisions as part of the collective of consumers who purchase and use these goods and services. These collectives of organizations and consumers make up the economy, the subsystem that encompasses the production consumption cycle depicted in the closed circular flow model. However, unlike the closed circular flow model, this model is open to the greater social system in which the economy is embedded and the greater ecosystem to which all humankind is bound. Thus, the model depicts the solar flow of energy that provides fuel for the earth's subsystems to operate, it depicts the terrestrial resources in nature that provide the materials and energy necessary for economic activity, and it depicts the wastes that are generated during economic activity that must somehow be absorbed back into the ecosystem. Nature provides the framework for this model. As with nature, waste is considered lost profit, and value is created by design. Feedback becomes the most valuable source of capital, and the more it is used, the more abundant it becomes. Limits are positive forces that encourage innovation and continuous improvement, and diversity brings choice, resilience, and sustainability. In other words, within this model, organizations can create more value by emulating what they once sought to conquer: nature.[2]

Strategic Managers: Critical Change Agents

The most powerful economic decision makers in the world are strategic managers in business organizations. These are the people who guide the world's economic machine, deciding what to produce, how to produce it, how to distribute it, where to distribute it, and so forth for all of the world's economic products and services. Collectively, their decisions in these matters contribute significantly to the structure and welfare of society as well as the amounts and types of energy used, resources consumed, wastes generated, and pollution created. Given this high level of influence that strategic managers have on the planet's social and environmental systems, it is critical that they take an active role in helping to institute the types of changes that will allow humankind to transcend into a community of economic, social, and ecological balance that can be sustained for generations to come.

In order to be effective change agents in humankind's search for a more sustainable system, strategic managers will need to firmly establish the assumptions and values of sustainable strategic management (such as those exemplified by Anthony Flaccavento and Bill Ford) into their organizations. This will require that they guide their organizations through shifts in thinking from the old closed circular flow paradigm to the new open living systems paradigm. At the heart of any paradigm shift are changes in the underlying assumptions and values that support the various paradigms. Clearly, the assumptions and values that underlie the closed circular flow paradigm (Figure 1.1) are quite different than those that underlie the open living systems paradigm (Figure 1.2). One assumes no limits to the economy, and the other assumes that there are both social and biophysical limits. One sees individuals, communities, and societies as value-free consumption-production machines, and the other sees individuals, communities, and societies as multidimensional, value-driven, living entities.

Transcending from the values and assumptions that underlie the closed circular flow view of the economy to the values and assumptions that underlie the open living systems view of the economy exemplifies the struggles that humans have experienced (and continue to experience) as they have morally and socially developed over the ages. Humans have struggled to rise above their basic individual instincts for survival, seeking harmony and safety with kin and others like themselves. They have sought meaning and purpose beyond their own tribal existence, and they have contemplated the greater universal powers that they do not understand. In recent times they have developed the ideas and processes of scientific inquiry and exploration, using them as paths to incredible economic and technological prosperity. All of these transitions in human development have been and continue to be fraught with doubt and conflict. The shift to the consciousness of sustainability, a shift that is already occurring in many pockets of society, is no different. Significant doubt and conflict have been and will be associated with humankind's transcendence from an individualistic, rational, scientific, objective society where nature and humans are considered usable and exploitable to a society that stresses human networks, communities, consensus building, diversity, and ecological balance.[3] This is the transition that strategic managers will have to lead their organizations through as they pursue sustainable strategic management. Such a transition will require that strategic managers guide their organizations through fundamental, transformational change processes that will instill new ways of thinking in their organizations regarding the relationship among economic success, social welfare, and ecological balance.

Transformational Change: Pathway to Sustainable Strategic Management

There are several basic requirements for effective planned change efforts in organizations. First, effective planned change efforts require the active support and participation of strategic managers in the change process. Second, effective planned change efforts need to be collaborative, team based, and participative, allowing for dialogue and consensus building with regard to what issues exist, what changes are to be made, and how those changes are made. Third, effective planned change efforts require that the changes be effectively integrated into the culture of the organization. Taken together, these requirements suggest that strategic managers wishing to achieve change in organizations must be willing to personally guide and participate in the change, they must be able to create and implement the open communication processes that will allow for broad meaningful dialogue in the change process throughout the organization, and they must successfully target the change toward the appropriate level of organizational culture in which the change has to occur.

An organization's culture is composed of the artifacts (language, rituals, symbols, architecture, etc.), norms, values, and assumptions shared by its members. These cultural elements exist in a hierarchy ranging from the more surface-level artifacts and norms to the deeper-level values and assumptions. That is, the essence of an organization's culture is defined by its fundamental shared values and assumptions, and this essence is reflected in and supported by the organization's shared norms and artifacts. There are varying degrees of organizational change depending on the level of organizational culture targeted by the change effort. Change efforts targeted at the shallower levels of the organization's culture (often referred to as first-order and second-order change) are adaptive and incremental, focusing on changing artifacts and norms in order that the organization can do better or do differently what it currently does. Such change is made precisely because the organization is seeking ways to better support current values and assumptions, so no change in fundamental values and assumptions is necessary. However, as change touches the more fundamental value-based levels of culture, change efforts become transformational (often referred to as third-order change), requiring organizations to employ generative learning processes that allow them to closely examine and change the underlying values and assumptions that define their essence. Transformational change is discontinuous, requiring that organizations achieve and perpetuate an entirely different qualitative state. Organizations attempting transformational change cannot expect to be successful by taking the slow linear steps associated with incremental organizational change efforts. Rather, transformational change requires fundamental efforts designed to completely shift the consciousness of the firm to a different level based on new values and new assumptions. Transformational change is radical, and it can be both expensive and disruptive.[4]

Scholars have long held that successfully implementing sustainable strategic management will require transformational change processes that allow organizations to examine and redefine their core assumptions and values as well as the nature of their work and the roles of their employees.[5] Post and Altman said in 1992, "Internal paradigm shifts and transformational change are necessary as companies attempt to adjust to the rapidly changing world of green politics and markets."[6] As social and ecological concerns continue to move into the forefront of the strategic issues facing business, adopting the values, assumptions, and processes of sustainable strategic management becomes more critical for organizations. We hope that this book serves as an informative guide for strategic managers and students of strategic management as they seek to meet this critical need.

Preview of the Book

In this book we will explore sustainable strategic management in depth. We present frameworks, concepts, and examples that we hope will help our readers to understand, formulate, implement, and evaluate sustainable strategic management in business organizations. In doing so, we hope to communicate to our readers that creating strategic management processes that strike a positive synergistic balance among economic wealth, social welfare, and ecological health is critically important, feasible, and potentially profitable, and we hope to provide our readers with some direction regarding how this can be done.

The book is divided into three parts. Part I, Foundations of Sustainable Strategic Management, is composed of the first three chapters. In Chapter 2, we discuss the concept of sustainability in depth; we present several scientific principles at the root of sustainability, we discuss the three dimensions of sustainability, and we discuss how achieving sustainability will require a change in the way humans think about their relationships with each other and with the planet. In Chapter 3, we present an in-depth introduction to the concept of sustainable strategic management. We will define the concept more thoroughly, discuss its basic elements, and examine the roles and responsibilities of CEOs and boards of directors in sustainable strategic management.

Part II of the book, Formulating Sustainable Strategic Management, includes Chapters 4 through 7. In Chapter 4, we examine the assessment of the organization's external environment as it relates to formulating sustainable strategic management strategies, focusing attention on the social and environmental issues facing business organizations, the relationship between sustainability and free trade, and coevolutionary industry analysis. In Chapter 5, we discuss internal assessment as it relates to sustainable strategic management. We discuss the need to modify value-chain analysis to account more fully for ecological and social concerns, and the need to include the earth as an organizational stakeholder with considerable clout in the formulation of organizational goals and strategies. In Chapter 6, we provide a general discussion of the types of strategies related to sustainable strategic management. We present functional level, competitive level, and corporate level sustainable strategic management strategies and their contributions to achieving eco- and socioefficiency and eco- and socioeffectiveness. In Chapter 7, we discuss the idea of strategic choice as it relates to formulating sustainable strategic management. We discuss the critical role that values play in the decisions made by strategic managers, and we present a set of values that supports sustainable strategic management.

Part III of the book, Implementing and Evaluating Sustainable Strategic

Management, includes Chapters 8 through 11. In Chapter 8, we examine the systems that need to be in place to support sustainable strategic management, including sustainable systems for research and development, procurement, operations, infrastructure, and marketing. In Chapter 9, we discuss the implementation of the cultures, human resource management practices, structures, and technologies necessary for successful sustainable strategic management. In Chapter 10, we discuss the evaluation of sustainable strategic management processes, focusing on sustainable accounting, finance, information systems, and reporting. We conclude the book in Chapter 11 by revisiting Anthony Flaccavento and Bill Ford as a means of assessing where we are today and asking where we may be tomorrow with regard to sustainable strategic management.

2

In Search of Sustainability

As we begin our journey into sustainable strategic management, we want to explore a basic question: Where will sustainable strategic management lead humankind? That is, if economic institutions were to universally adopt sustainable strategic management processes, what would the potential outcome be? The short answer to this question is that sustainable strategic management will contribute to humankind's pursuit of *sustainability*, the point at which humankind "meets the needs of the present without compromising the ability of the future generations to meet their own needs."[1] But what does this mean? What are the underlying principles and dimensions of sustainability, and what will be required for sustainability to happen?

In an effort to address these questions, we will examine the concept of sustainability in some depth in this chapter. In doing so, we are establishing sustainability as both an ideal end state for a world dominated by business organizations pursuing the values, principles, and practices of sustainable strategic management, and a viable transformational process that can help humankind to achieve this end state. However, as we proceed we want to advise the readers that sustainability is a complex, transdisciplinary, multidimensional concept. It is not something that humankind is close to achieving or even fully understanding. As such, achieving sustainability will require that humans think differently—not just act differently—about the relationships between economics, society, and the natural environment.

The Science of Sustainability

We would like to begin our discussion by briefly presenting some of the basic scientific concepts that help ground sustainability in well-established scientific frameworks with regard to the relationships among economic activity, society, and nature. We contend in this section that the earth is a living system, that the relationships among the earth's components are coevolutionary, that economic activity on earth is subject to the laws of thermodynamics, and that economic activity on earth is metabolic in nature.

The Earth Is a Living System

The term "living system" refers to something that exhibits the characteristics of life.[2] Living systems do not have to be alive in the biological sense, but they do have biological functions such as birth, death, and reproduction. Living systems are open, meaning they exchange information, matter, and energy with their environments in order to counteract uncertainty and decay. They receive feedback from their environments that helps them balance inputs and outputs to maintain a dynamic equilibrium. Living systems are morphogenetic, meaning that they can renew, reproduce, or regenerate themselves. Living systems have some purpose, goal, or final state that they seek. Living systems are complex; they are composed of a finite number of component subsystems, each of which processes, information, matter, or energy. These component subsystems are highly interdependent and cannot be treated as isolated entities. Living systems are irreducible wholes; their survival is threatened when their component subsystems break down. Further, living systems are synergistic, displaying certain properties that could never be anticipated by analyzing their component subsystems; they are different from the sum of their parts. The component subsystems of living systems exist in nested hierarchies, with each subsystem contained in a larger system. These hierarchies of subsystems exist at different levels of physical, social, or spiritual complexity. Each level both includes and transcends the previous level.

As a living system, the earth both encompasses and transcends the matter, plants, animals, people, societies, and organizations that compose it. From its environment it imports sunlight, which provides the energy for resource development and life itself. The planet's survival depends on the delicate interaction among the atmosphere, oceans, land, species, and other subsystems that compose it. Figure 1.1 (Chapter 1) represents a nested hierarchy of living subsystems associated with economic activity on earth. As such, all of these subsystems—individuals, organizations, the economy, society, and the ecosystem—are complex, morphogenetic, and interdependent; all exchange information, energy, and matter with their environments to survive; and all are interdependent with one another. Each subsystem is both composed of and qualitatively different from those below it, and the demise of any of these subsystems would gravely threaten the survival of the others. Thus, achieving sustainability in any of these subsystems means achieving a sustainable balance among all of them.

The Earth's Subsystems Coevolve

A useful framework for understanding the dynamics of sustainability is Gaia theory.[3] Gaia theorists point out that the living earth was actually born with the big bang of creation some 15 billion years ago, when the energy necessary for the formation of the universe was released. Some 4.5 billion years ago the earth became a discernable hot, gaseous ball, and since then it has changed dramatically in terms of chemical content, geological activity, and the evolution of life. Research in Gaia theory has demonstrated that these three factors have not changed separately, but rather have coevolved on the planet. For example, the oxygen content of the atmosphere has increased over the eons to 21 percent—enough to support large mammals such as humans—as a result of interactions among the earth's living and nonliving components. In this and many other ways, the earth's living organisms have continuously interacted with their natural environment to change and regulate chemical, atmospheric, and climatic processes in much the same way that a plant or animal self-regulates its internal state.

These symbiotic, coevolutionary relationships are seldom simple, generally involving a complex choreography of both cooperation and competition. Thus, Gaia theory has clearly established that the relationship between the earth's biological and physical forces is one of mutual influence. Gaia theory also points to the fact that humankind's environmental sensitivity need not be altruistic. Although environmental debates are often couched in terms of "saving the planet," research results from Gaia theorists make it clear that the planet can take care of itself. What is threatened via ecological and social degradation is not the planet but humankind and its way of life. Thus, achieving sustainability will require balanced, complex interactions involving both cooperation and competition among all of the planet's subsystems, or the human condition will suffer as a result.

The Economic Subsystem Is Entropic

As mentioned, survival for living systems is based on the ability of those systems to exchange energy, wastes, and information with their environment. It is a process of constantly swimming upstream against time, seeking order in a sea of chaos. This process is subject to the laws of thermodynamics, a set of principles governing the movement and transformation of energy in the universe.[4] The first law of thermodynamics, the conservation law, says that the amount of energy released by the big bang is a constant in the universe. Energy cannot be created nor destroyed; it can only be transformed from one state to another. The amount of energy generated during this transformation depends on the temperature difference between the states (hence the term "thermodynamics"). The second law of thermodynamics says that every time energy is transformed from one state to another, some of its available energy to do work is lost. This process is called "entropy." Entropy occurs when stored energy becomes cooler, less concentrated, or less ordered when it is applied to do work. When energy is no longer available to do work, when it has degraded to the point of being useless, it becomes waste.

Whereas entropy is a certainty for the earth, there is little certainty about the path or time it will take. These will depend on how efficiently humankind uses its available energy and how well it responds to the changes in its environment. The earth and its living subsystems can survive and increase in orderliness while there is sufficient power from the sun as long as people respond correctly to signals from the environment. Global warming, smog, cancer, water shortages, genocide, and energy crises are just a few signals that indicate the need for changes in how humans interact with the planet. The more serious these problems get, the more difficult they will be to deal with. However, if people respond appropriately to these signals, the species can survive and develop for eons to come.

The path that entropy will take is directly related to economic activity on the planet. To this point, humankind's economic system has survived by using processes that rapidly transform energy and natural resources from their low-entropy natural state to create high-entropy products, services, and wastes. In the economic system, money and energy flow in opposite directions. For example, the farmers' money goes to town in exchange for the fertilizer they need to power their crops; the manufacturers' money goes to the utility company in exchange for the power they need to produce their products. However, while money stays within the economic system, energy often exists outside the system. Sunlight, nonrenewable resources, and other sources of energy that power the economy do not enter the economic cycle until they are purchased or converted to fuel. Further, the wastes that occur as a result of converting energy into economic wealth are also considered external to the system. Thus, it seems reasonable to assume that the entropy law should be at the heart of future economic theory and practice. If it is not, it is virtually impossible to effectively account for the true value of natural resources, the intrinsic value of life, and the actual cost of pollution and overpopulation in economic activity. Essentially, assuming that economic activity is not subject to the entropy law leads directly to the fallacious assumption that unlimited economic expansion is forever possible.

The Economic Subsystem Is Metabolic

One of the most enlightening frameworks for understanding the impact of a high-entropy economy is *industrial metabolism*.[5] Just as living organisms have metabolic processes for transforming the energy they import from their environment into life-maintaining processes, economies can also be viewed as metabolic because they extract large quantities of energy-rich matter from the environment and transform it into products for consumption. Industrial metabolism involves all the processes used to convert resources, energy, and labor into products, services, and wastes.

Whereas the metabolic processes necessary to maintain life in the ecosystem are balanced and self-sustaining, metabolism in the economic system is grossly out of balance with its environment. Resources that literally take eons to renew (such as oil) are being used at nonrenewable rates because only a small percentage of the resources used in economic activity remains in the system for any length of time (basically as durable goods). Most materials are used to produce food, fuel, and throwaway products that pass through the economic system from extraction to production to consumption to waste very rapidly. These wastes are often toxic and harmful to the natural environment. The damage is done not within the economic system per se, but in the atmosphere, water, and gene pool that have no current economic value.

Entropy occurs at all points in the metabolic process, including extraction, production, and consumption. However, most of the loss comes at the point of consumption. Most foods, fuels, paper, lubricants, solvents, fertilizers, pesticides, cosmetics, pharmaceuticals, and toxic heavy metals are discarded as wastes after a single use, as are thousands of other products. Many of these are very difficult and expensive to recycle, so people not only use too many of them but also are not likely to use them again.

The basic message from the industrial metabolism framework is that the metabolic processes in the economy need to achieve the same type of balance that is possible in the ecosystem when it is absent of economic activity. Just as the ecosystem can sustain itself indefinitely by importing sunlight and using it to power a system that operates almost totally by recycling materials, economic systems also need to incorporate sustainable energy transformation processes. Achieving this balance will require total materials recycling. That is, the four segments of the materials flow cycle—the natural environment, raw materials and commodities, productive capital, and final products—must achieve a balance via processes such as recycling, remanufacturing, reconditioning, and so forth.

Dimensions of Sustainability

As we mentioned above, sustainability is a complex concept that is far from being fully developed and understood. To this point it is commonly held among scholars and practitioners alike that sustainability encompasses three broad dimensions: economic, social, and ecological.[6] A key to understanding these dimensions is to recognize that they are highly interdependent. For example, population growth is a serious social problem with huge ecological and economic implications, and climate change is directly related to economic activity and has potentially serious social consequences. Below we discuss these three dimensions in some depth, focusing on both where they stand currently and in what direction humankind needs to take them in order to achieve sustainability.

The Economic Dimension

Post–World War II global economic growth has been phenomenal. Since 1950, real gross world product has multiplied almost sevenfold, growing from $6.6 trillion to $45.9 trillion. This translates into an almost threefold increase in gross world product per person, from $2,582 to $7,454. Further, since 1970, the global economy has soared with global exports rising from $1.45 trillion to $7.43 trillion.

Unfortunately, as positive as these figures seem, sustainability cannot be achieved unless understanding of the economic dimension goes beyond the current thinking that economic growth is the only important measure of economic well-being. The economic dimension of sustainability also involves the need to create for posterity an ecologically balanced and socially just economic system that provides humans with the goods, services, economic justice, and meaningful employment necessary for a high quality of life. As we discussed above, the rapid entropy associated with high economic growth rates is not sustainable over the long run. In addition, despite the phenomenal growth rates, roughly 20 percent of the global population lives in extreme poverty, surviving on less than $1 per day. Thus, at the heart of sustainable strategic management is the understanding that the impacts of economic activity should be measured on their ecological and social as well as economic outcomes. As we will discuss in some depth later in the book, this will require that organizations go beyond relying on traditional economic inputs and financial analysis processes.

The Social Dimension

Socially, sustainability refers to the need to effectively deal with a plethora of social issues facing communities and nations worldwide. Included among these issues are population growth, the economic gulf between the developed and developing worlds, human rights, human health, gender equity, the digital divide, and community viability.

There were 250 million people on earth 2000 years ago, and only 500 million people on earth in 1650—it took over a millennium and a half for the population to double. The population doubled again to 1 billion by 1850. From 1850 to 1987 the population multiplied fivefold to 5 billion, and between 1987 and 1999 another 1 billion souls were added. The planet now houses 6.2 billion people and is currently growing at an annual rate of 1.1 percent. An additional 77 million people—the population of Germany—were added to the planet in 2001 alone. Ninety-five percent of this growth is taking place in developing nations, with sub-Saharan Africa leading the way. Rapid population growth exacerbates myriad economic and social issues, such as poverty, discrimination, violence against women, infant mortality, and inadequate reproductive health care. Rapid population growth also puts major stress on natural resources, including water, forests, cropland, and species diversity. These complex, highly interrelated issues make improving the quality of life of citizens in nations and regions experiencing rapid population growth a tough uphill battle.

Population growth patterns are a major contributor to the ecological, economic, and social inequities that currently exist between the developed and developing nations of the world. The 25 percent or so of the earth's population living in the developed world (mostly in the northern hemisphere) control most of the world's financial resources, consume most of its goods and services, and use most of the natural resources extracted from the planet. The 75 percent living in the developing world (primarily in the southern hemisphere but also including indigenous peoples and other minorities all over the globe) get fewer benefits from the global economic system, often having to live on scarce resources without meeting their basic needs. This economic divide manifests itself in numerous human rights issues, including health-care and literacy gaps, gender inequities, and a digital divide.

There is a huge health-care gap between the developed and developing nations. The United States spends over $4,000 per year per person on health care, but India spends only $73 per person. The United States has an infant mortality rate of 4 per 1,000 live births while India's rate is 70 per 1,000. Infectious diseases such as cholera and tuberculosis are rampant in developing nations. The same is true for AIDS; for example, 20 percent of the people of South Africa and 36 percent of the people of Botswana are infected with the HIV virus.

Another developed-developing nation gap exists in literacy rates. Whereas the developed world can boast that almost all of its adult citizens are literate, the illiteracy rates in many developing nations are alarmingly high; for example, in India 45 percent of the adult males and 57 percent of the adult females are illiterate. It is also important to note that literacy rates in developed nations often vary significantly with levels of income. For example, illiteracy is a serious problem in the Southern Appalachian Mountains of the United States, one of the nation's most poverty-prone regions. Adult functional illiteracy rates of 30 percent to 50 percent are not uncommon in rural areas of this region.

One of the key social dimensions of the population issue is gender inequity. Undervaluing, abusing, and otherwise discriminating against women is a worldwide problem. In many cultures women are seen as inferior persons whose primary purpose is to serve as vessels for male sexual pleasure and childbearing. Women brave enough to acquire birth control pills in developing nations often have to hide them from their husbands or lovers for fear that they will be accused of interfering with the men's right to control their fertility. Levels of income, education, legal rights, and political involvement are lower for women than for men throughout the world. One of the most egregious outcomes of gender inequity is violence against women, whether it is verbal, physical, sexual, or economic.

Another huge gap between developed and developing nations and rich and poor people is the digital divide.[7] One of the most significant social trends of the past twenty years has been the meteoric rise in both the quality and use of information and communication technology. Electronic access to information has exploded worldwide. Unfortunately, the explosion has benefited the developed world much more than it has the developing world. For example, whereas worldwide only about 0.1 percent of the people who earn less than $1,000 per year have access to the Internet, some 60 percent of those earning over $40,000 per year have access. Given the phenomenal growth in electronically based economic activity, gaps such as this sorely threaten the future of development in regions with poor access to information and communication technology, and left unattended the digital divide may very well contribute to an ever widening social gap between the developed and developing world. Envy related to the digital divide explains why many citizens in the Philippines celebrated when a Philippine student was able to crash the computers of some of the world's largest economic institutions with the "love bug" virus in 2000.

At the heart of improving most of the social issues mentioned above is finding ways to control population growth. China, of course, has instituted a one-child-per-family policy, but such policies would be difficult if not impossible to implement in freer, less rigidly controlled nations. However, there is good evidence to suggest that free societies that stress gender equity in education, legal rights, and economic standing tend to experience reductions in their population growth rates without such control-oriented policies. There is also evidence to suggest that controlling population growth is achievable via improved education, improved health care, and improved economic opportunities for all citizens, not just women. Improved educational opportunities and economic development are also keys to closing the digital divide. In this regard, improved information infrastructure and supportive government policies and institutions are important as well.

A social issue related to the rapid expansion and globalization of economic activity is the viability of communities. At the heart of this issue are two dimensions. The first dimension is the depletion of natural resources that fuel a community's economy. The decline of dozens of formerly thriving fishing villages in the northeastern United States because of overfishing in the Georges Banks clearly demonstrates how resource depletion can impact communities. The second dimension is the loss of the economic base of a community because of the relocation of global corporations seeking cheaper labor and materials, less stringent environmental regulations, and so forth.[8] This was a problem in Johnson County, Tennessee, located in the Southern Appalachian region of the United States. This small rural community has been plagued by economic collapse caused by losing three large employers, Levi-Straus, Sara Lee Clothing, and Timberland Shoes, to new locations abroad. As a result, unemployment rates are very high in the community, educational spending (which was already very low) is plummeting, and citizens are forced to seek jobs many miles from their homes. Solutions to these community viability issues are complex and multidimensional. Such solutions may involve implementing interlocking community-based decision-making structures that protect individuals' rights to make choices while allowing for coordination and involvement in decisions that affect the general social and environmental health of the community. For example, many organizations today develop community councils composed of organizational members, environmentalists, government officials, and social activists designed to serve as information-sharing and action channels with regard to the organization's decisions and activities. Another important approach to these community viability issues involves creating locally based economic systems that are less dependent on global corporations, distant energy sources, and so forth.

The Ecological Dimension

As population and economic activity expand, the production and consumption of goods and services continue to grow worldwide. From this ever-expanding cycle of production and consumption come unsustainable levels of resource depletion and waste generation. This in turn results in the increasing types and severity of environmental ills facing the planet. Metals production is up 45 percent in the past thirty years. Some 2.4 billion people currently live in water-stressed nations, and that figure will likely increase by about 50 percent over the next twenty-five years. Somewhere between 10 percent and 20 percent of the world's cropland is degraded because of soil erosion, salinization, and so forth. There are between 300 million and 500 million tons of hazardous wastes generated and disposed of each year worldwide. Global carbon emissions increased approximately 9 percent in the 1990s, reaching 68 billion tons per year by 2000. These are just a few of the patterns indicating that current human activity is ecologically unsustainable. Thus, from an ecological perspective, sustainability refers to the need to limit the amount of production, consumption, resource depletion, and waste generation to levels that the planet can realistically absorb without threatening the atmospheric, geospheric, and biospheric processes that support human life for posterity.

As discussed, one key to effectively dealing with such issues is to bring industrial metabolism into kilter with the planet's natural processes via total materials recycling. If this can be achieved, it will contribute to an industrial system with extraction rates that are sustainable over time, and it will contribute to the generation of both the types and quantities of wastes that are readily absorbable by nature. Very much interrelated with total materials recycling is the need to make a transition to a sustainable worldwide energy system. Contrary to popular belief, the earth has an abundance of energy. The issue with energy is not the quantity of the energy but its sources and generation processes. Humankind's current dependence on fossil fuels and nuclear fission is not sustainable because of the environmentally degrading nature of these energy sources. Developing a sustainable energy system will require developing a worldwide mix of energy sources and technologies that are more renewable, less dissipative, and more ecologically sensitive than the current fossil-fuel and nuclear-based systems. The transition to such a system will be neither easy nor cheap. It will require continued investments in improved storage technologies for solar- and wind-generated power, improved solar electric-generating technologies, improved wind turbines, and so forth. However, as investments in these technologies continue to grow, these energy sources can become more economically competitive. Both wind and solar energy use are currently increasing at over 30 percent per year, which is much faster than growth in fossil fuel and nuclear energy sources.

Achieving a system that accounts for these material and energy concerns will require transcending three progressively difficult stages of industrial evolution.[9] The first stage, the type I industrial ecosystem, is the classical industrial model. In this stage the global production and distribution systems operate on straight linear processes in which virgin raw materials and energy are converted into goods and services. The byproducts of this process are heat and material wastes that either dissipate or must be disposed of in the natural environment. The second stage of industrial evolution, a type II industrial ecosystem, involves some recycling of materials and energy in production processes, but still requires the linear transformation of virgin inputs and energy into products and wastes that must be absorbed by nature. The third stage of evolution is to a type III industrial ecosystem in which the only inputs are renewable energy, and operations are totally closed-looped with virtually total materials reuse and recycling. Type III industrial ecosystems export only heat into the external environment. These systems mimic mature natural ecosystems that are generally quite stable, operating on minimal amounts of entropy. Today industrial systems seem to be in transition from type I to type II systems. We believe that sustainable strategic management can provide pathways to type III systems.

Sustainability: A New Way of Thinking

As demonstrated in the above discussion, there are many barriers to achieving sustainability, and they are complex, incredibly intertwined, serious, and, some say, overwhelming. At the biophysical level, the entropy law provides the absolute physical wall beyond which human activity on earth will not be possible. In order to bring the economic system into sync with the earth's natural entropic processes, humankind must find ways to slow down the high-entropy energy, resource, and waste processes that result from current economic activities. This means overcoming several biophysical barriers, including: finding safe and plentiful substitutes for the nonrenewable resources and toxic chemicals now in use, developing the efficient use of clean, renewable energy sources, developing better processes for recovering, recycling, and disposing of wastes, developing more efficient production processes, and developing closed-loop networks of industrial ecosystems throughout the globe.

Sustainability is not just a biophysical problem. If it were, then maybe it would be easier to confront. However, as we discussed, sustainability is a human development problem, with all of the ethical, cultural, social, religious, political, civil, and legal implications that entails. Slowing population growth to sustainable levels will require addressing an array of issues including gender equity, economic equity, health care, education, birth control, social mores, and religious principles. Curtailing slash-and-burn destruction of the rainforests will not stop without addressing issues like democracy, human rights, property rights, and international trade. Solving the problems of the unsustainable megalopolises around the globe means addressing all these same issues and many others, such as how to finance sustainable public transportation systems and how to find employment for the millions of poor who migrate to these cities every year. To add to the complexity is the fact that these are not just the problems of the current generation; they are the problems of those not yet living. They are the problems of our children, and our children's children. They are also the problems of the other species on the planet, both current and future.

When viewed from the lens of rational thought and scientific inquiry that has dominated the way humans have viewed the universe for the past 350 years or so, the odds of coming up with solutions to the problems related to the interface among business, society, and the natural environment seem slim at best. The idea that Adam Smith's *economic man* will ever willingly surrender his quest for castles and gold so that people he will never know can have a comfortable place to live in a safe society with clean air to breathe, adequate soil for food, clean water to drink, and the opportunity for creative self-expression is essentially ludicrous within the current framework of the materialistic, ego-centered, growth-oriented, mechanical, mental models that are currently driving human thought processes.

In short, we believe that achieving sustainability is not just about changing how humans do things; it is about changing how humans view things. If humans are ever to see the light of sustainability, it will not be because they were simply rational, logical, and scientific; it will be because, in addition to these things, they were able to change the underlying mental frameworks that guide the way they see the planet and their place on it.

Many terms are used to describe the underlying mental frameworks that guide human thought and action, including myths, archetypes, and paradigms. Regardless of the moniker, these are the basic mental models upon which humans base their interpretations of truth. These mental frameworks are in essence beyond truth or falsehood. Rather, they are the metaphysical frameworks within which people determine truth and falsehood. In this sense, they are true when they guide individuals and societies on paths of living that are in harmony with themselves, with other humans, and with nature, and they are false when their guidance leads to conflicting, confrontational relationships with themselves, other humans, and nature. As such, these myths, archetypes, and paradigms are the underlying mental models that define how humans instruct their young, how humans are expected to mature, the roles humans are expected to play in society, the relationships among humans, society, and nature, and the paths to human and societal spiritual fulfillment.

Mythologist and anthropologist Joseph Campbell said that you could tell the dominant myth of a given society by examining the height of its buildings.[10] He said that the multistory seats of economic activity that define the skylines of our cities today demonstrate that humankind's most dominant current myth is one of economic wealth. The terrorist attacks on the World Trade Center on September 11, 2001, certainly add credence to Campbell's contention, demonstrating in a few tragic moments both the power and the vulnerability of a dominant mental model.

Unfortunately, as we tried to demonstrate in our discussion of the economic, social, and ecological dimensions of sustainability, the dominance of the economic wealth myth is not meeting the acid test of truth; it is not currently leading humankind down the path of harmony and concord with one another and with nature. Indications are that humankind needs a new way of thinking that will put it on such a path, and many believe that sustainability is at the heart of this new way of thinking.

Pioneers of Sustainability Thinking

Many prominent scholars agree with Joseph Campbell, arguing that the current economic wealth framework grew out of the Scientific and Industrial Revolutions that began in the seventeenth century and has since come to dominate all of society's institutions, whether they are political, religious, educational, or economic.[11] They believe that this view of the world—rooted in a Cartesian, Newtonian view of a quantitative, clockwork universe in which humans are only objective observers—is both inaccurate and inadequate for humankind's survival. They contend that the assumptions of the current mental framework are based primarily on the concepts of objectivity, rationality, dominance, and cause-effect, with no meaningful accounting for quality, aesthetics, social ills, and the mutual causality of nature. Two of the true pioneers that have championed this new way of thinking are Aldo Leopold and E.F. Schumacher. Below we summarize some of their contributions.

Aldo Leopold: Voice of the Land

Aldo Leopold (1887–1948) was a renowned environmental scholar and spiritual leader of the wildlife conservation movement in America, and his work is considered by many to be at the center of modern environmental ethics.[12] Leopold was educated in forestry at Yale University and served in the U.S. Forest Service in the southwestern United States. However, for most of his career he was a forestry professor at the University of Wisconsin.

In commenting on the need to change the way we think about the relationships between humans, society, and nature, Leopold said, "We abuse land because we regard it as a commodity belonging to us. When we see land as a community to which we belong, we may begin to use it with love and respect."[13] Leopold tied his ideas about the role of ecology in human history back thousands of years to the biblical teachings of Abraham, Ezekiel, and Isaiah. He contended that humankind is in its current ecological predicament because it has ignored these teachings, focusing instead on the human actors who conquered nature and other humans in the process of establishing the human community. He provided clear instruction on the interconnectedness of nature, the role of land in economics and politics, and the fallacies of the assumption that nature is humankind's personal possession.

Leopold suggested that humans adopt a new mental model based on what he calls the *land ethic*, and, true to his training as an ecologist, he chose the concept of community on which to rest the development of this idea. He did so because he believed that the membership of individuals in communities tempers their instinct to compete with their instinct to cooperate, improving their potential to coevolve effectively with their environment over time.

Leopold pointed out that the ethical standards of humankind have evolved through two stages with the evolution from stage to stage being characterized by humankind's collective decision to redefine the concept of property. Stage one was a human-human ethic. This system of ethics extended ethical considerations to certain persons, such as spouses, friends, kin, and the economically and politically powerful, but withheld it from others. Under it, slavery and indentured servitude were tolerated, classifying these people as mere property not worthy of ethical consideration. Stage two widened the focus of ethics from human-human to human-society; Leopold cited democracy as a system that emerged out of this evolution. Under a human-society ethic, slavery is not tolerated; all people fall within the venue of ethical rights and responsibilities. Leopold said that it is now time for the transition to the human-land stage of ethical evolution, in which the status of the land is raised from property to a full member of the human community. He said, "The land-relation is still strictly economic, entailing privilege but not obligations. The extension of ethics to this third element in the human environment is . . . an evolutionary possibility and an ecological necessity."[14]

Leopold pointed out that humankind could not completely evolve to a land ethic without a shift in human consciousness. He believed that conservation efforts in the late 1800s and early 1900s were failures because, although they stressed the need to educate the public about the natural environment, they ignored the need to change the public's basic perceptions about its relationship to nature. He stressed that "a land ethic reflects the existence of an ecological conscience,"[15] and he suggested that for such a consciousness to emerge, humans have to collectively perceive nature as an interconnected web composed of the planet's organisms and processes. He suggested that the biotic pyramid provides a valid image for this new consciousness because it demonstrates the energy circuits connecting the sun to the soil, the soil to the plants, and so on up the layers of the food chain. This circuitry is totally interconnected, meaning that change in any part of the circuits requires that the other parts adjust themselves. He said that creating an image based on the biotic pyramid can have three significant effects on human consciousness, which, in turn, can facilitate the evolution of a sustainability-based mental model on the planet. First, it will create awareness that land is something other than soil. Second, it will create an understanding that plants and animals keep the energy circuits necessary for human existence open. Third, it will create an understanding that recent human changes to the pyramid are qualitatively different from those that occur in nature, with more far-reaching results than anticipated.

E.F. Schumacher: Guiding the Perplexed

E.F. Schumacher (1911–1977), an erudite gentleman, was a brilliant economist who attended Oxford and Columbia.[16] Lord Keynes once deemed Schumacher the most worthy candidate to succeed him as the world's top economist. As is true for most people, Schumacher's ideas matured and changed as he matured and changed. He began life as an arrogant, atheistic intellectual who disdained formal education and religion because he believed himself smarter than his teachers and ministers. He focused only on facts, believing that they explained everything. However, as he grew older he turned to Buddhism and eventually to Catholicism. As his religious and social enlightenment progressed, his ideas about economic systems became more spiritual than intellectual, and he found that the most important things in life were learned from the common man and from the soil, not from the economic experts with all their facts and figures.

Like Leopold, Schumacher believed that humankind's ethical sphere needed to be broadened to encompass the entire planet and all of its species. Schumacher based his philosophy on the concept of *higher*. He used the progression of earth's inhabitants, minerals–plants–animals–humans, to demonstrate that reality is at once empirical and spiritual. He said that the difference between the lower-level minerals and higher-level plants is life, the difference between plants and higher-level animals is consciousness, and the difference between animals and higher-level humans is self-awareness. Thus, progression from lower to higher levels of understanding is a progression from outer to inner, from mechanistic to organic, from control to understanding, and from the head to the heart.

Schumacher was rather vocal in his criticisms of the influence of Cartesian philosophy on the ethics of the Industrial Age. Prior to the Cartesian influence, human mental frameworks were based on a firm belief that human happiness was achieved via a search for higher spiritual meaning and experience. However, this search was eroded by the philosophy of René Descartes (1596–1650). Descartes believed that humans' ability to reason placed them in a position separate from and above the environment in which they existed. Schumacher said that only the lifeless, lower mineral level lends itself to the Cartesian idea of a quantifiable universe. The objective, quantitative mental models that emerged from a Cartesian world, in Schumacher's opinion, created a utilitarian society in which value is believed to be a quantifiable concept. With the Cartesian paradigm came the belief among most people that nothing can exist beyond what can be observed and measured. With this came the ideas of Immanuel Kant (1724–1804) and others that humans do not really know what they want, what makes them happy, and what fulfills them because they do not have all of the facts. This led to the separation of spiritual life from secular life, which, in turn, resulted in the denigration of the value of spiritual experience and the glorification of material wealth.

According to Schumacher, within the philosophies of Descartes there is no room for including nature in humankind's critical mental sphere because the aesthetic value of the natural environment necessary for such inclusion is absent from the Cartesian system. Schumacher said that invisible powers such as life, consciousness, and self-awareness can be understood only by moving beyond quantifiable information to a higher level of knowledge. He believed that humans' happiness can be achieved only by developing their higher faculties, which he describes as the search for God, and that if they move lower, they will develop their lower faculties and make themselves deeply destructive and unhappy.

Schumacher, considered a statistical genius early in his career, believed that lower-level, quantitative measures of reality are quite sufficient for the plethora of linear, *convergent* problems that humankind faces regularly. Via the application of linear mathematics and rational thinking, convergent problems can be approached from a variety of directions that yield reasonable, remarkably similar answers. Schumacher pointed to the modes of transportation that have developed over the centuries as examples of the power of applying quantitative analysis and rational thought to convergent problems. However, he pointed out that problems related to higher-level, aesthetic values and ethics are not convergent in nature; rather, they are *divergent*. Unlike convergent problems, divergent problems defy solutions attained with rational linear logic. The more straight-line logic applied to such problems, the more diametrically opposed and outrageous the solutions become. Finding meaningful solutions to divergent problems means moving to a higher, transcendent level. Schumacher used the slogan of the French Revolution, "Liberty, equality, fraternity," to demonstrate this. The slogan represents two diametrically opposed, animal-level forces—liberty and equality—that can exist together only under the transcendent, human-level umbrella of fraternity. That is, only when a society's members have brotherly love and respect for one another can they be both free and equal.

The divergent problem focused on most extensively in the writings of Schumacher is the dichotomous relationship between the economic system and the natural environment. He said that modern industrial society is living under three dangerous assumptions: unlimited growth is possible in a finite world; there are unlimited numbers of people willing to perform mindless work for modest salaries; and science can be used to solve social and ecological problems. To Schumacher, these Cartesian assumptions are paths to resource depletion, environmental degradation, worker alienation, and violence. He coined the phrase "small is beautiful" to symbolize his belief that humankind's solutions to these problems lie in transcending to a higher level of analysis that can properly account for the aesthetics of both economy and ecology. At the heart of achieving that transcendence is sustainability.

Making the Mental Shift to Sustainability

Schumacher's famous saying, "small is beautiful," was meant as a challenge to the way humans think about their relationships with each other and with nature. The slogan challenges humankind to give up the bigger-is-better, grow-forever mindset that dominates the current materialistic world in favor of a belief that the long-run survival of the human species on the planet depends on finding for posterity a sustainable balance among the planetary economic, social, and ecological subsystems.

The challenges of integrating sustainability into humankind's mental fabric are twofold: The first challenge is to develop an understanding of sustainability that is sufficient to allow the concept to serve as a meaningful ideal end state upon which to base humankind's future. The second challenge is to find effective means for actually making the transformation to a sustainable society.

The magnitude of these challenges can be illustrated through an examination of the nature of the mental shifts required to adopt sustainability as both an ideal and a viable transformational process.[17] Under the economic wealth framework, the following assumptions are considered to define truth: The earth is passive, inert, mechanical, infinitely divisible, and legitimately exploitable. Humans are separate from and superior to their natural environment. The earth is infinite with an inexhaustible supply of resources. Science is the only proper framework within which to manage humankind's relationship with nature. Cost-benefit analysis is the appropriate tool for making decisions about the potential for human suffering and environmental harm. Humans will always be able to find technological solutions to their problems. Natural capital is a near perfect substitute for other forms of capital. The economy is a closed circular-flow of goods and services between households and organizations and is isolated from nature. Humans are self-serving, one-dimensional beings who pursue fulfillment solely by attempting to satisfy their unlimited, secular material wants. Social welfare is most easily improved via unlimited economic growth that allows for the continuous trickling down of wealth to the masses, meaning that poverty will eventually disappear if the rich can just get rich enough.

By contrast, a sustainability-based mental model of the world would define truth using very different assumptions, such as the following: The planet is home to humankind, and as such it should be kept clean, healthy, safe, and well managed. A healthy economy and a healthy ecosystem are intricately and irreversibly interconnected. Humans are both a part of nature, and, because of their superior intelligence, the chosen stewards of nature. Humans have ethical responsibilities that are intragenerational, intergenerational, and interspecies. The ecosystem is finite with limited resources and limited regenerative and assimilative powers. Irreplaceable parts of the natural capital, like other species and the ozone layer, are nonsubstitutable. Unlimited economic growth forever is impossible because the laws of thermodynamics govern interactions between the ecosystem and the economic system. The precautionary principle is the appropriate tool for assessing the potential for human suffering and ecological catastrophe. Humans are multidimensional beings who can learn to appreciate aesthetic as well as economic value and can learn that wisdom, intellectual development, and spiritual fulfillment are as important as material well-being. The economic system should internalize all ecological and social externalities, such as energy/matter throughput and poverty.

It should be clear from examining the deep differences between the economic-wealth and sustainability mental models described above that achieving sustainability is going to require a profound mental transformation. With such a transformation, sustainability will become embedded in the mythological conscience of humankind and serve as a guiding force for future thought and action.

Conclusions

We would like to end this chapter with an ancient story. There once was a boy who built a trap and captured a beautiful bird so that the bird would be able to sing only for him. The boy's father got angry with the boy, saying that the family could not afford to feed the bird just for its song. That night after everyone was asleep the father took the bird to the forest and killed it, but when the bird died, so did the father. The story is a metaphor for the unsustainable path that humankind has been going down since the dawn of the Industrial Revolution. Just as the boy used his superior brain and technology to capture the bird, humans have used their superior brains and technology to harness nature for their own ends. Just as the father made an economic decision to kill the bird, humans have been destroying the natural beauty and resources of the planet for short-term economic gains. And just as the death of the father followed the death of the bird, humans are risking the long-term survival of their species by destroying the natural environment that supports human life.

Sustainability provides an excellent framework for reversing the destructive trends implied in this story. We have attempted in this chapter to provide a discussion of sustainability that would provide some depth regarding what it means to actually "meet the needs of the present without compromising the ability of the future generations to meet their own needs." In doing so, we have emphasized the underlying principles and dimensions of sustainability that may serve as ideals to define human existence in the future, we have emphasized some of the key transformational processes related to pursuing sustainability, and we have emphasized that for humankind to ever fully embrace the ideal of sustainability and begin a full-scale transformation toward it, a shift in how humans view the world and how the world functions will be necessary.

Given the prominence of economic activity in the sustainability formula, there is no doubt that the business organizations that design, develop, produce, assemble, transport, sell, and dispose of the goods and services in the global marketplace are central to the transformation to a sustainable world. Whether these organizations continue to be part of the sustainability problem or become a part of the sustainability solution will rest in large part on the decisions the strategic managers in these organizations make. Therefore, strategic decision-making processes in business organizations are the keys to solving the puzzle of global sustainability. Thus, in the next chapter we will discuss the integration of sustainability into strategic management and the importance of the role of strategic managers in the process.

3

Roles and Responsibilities of Sustainable Strategic Managers

We defined sustainable strategic management (SSM) in Chapter 1 as strategic management processes that are economically competitive, socially responsible, and in balance with the cycles of nature. Like all long-term strategic initiatives, successfully implementing sustainable strategic management requires that the core philosophies and value systems of the organization be consistent with the initiative, and it requires that the organization create the appropriate strategies, capabilities, structures, and processes necessary for effective implementation. Successful sustainable strategic management also requires developing myriad internal and external economic, social, and environmental alliances, networks, and relationships with other firms, governments, interest groups, communities, activists, and so forth. Implementing sustainable strategic management requires crafting sustainable SSM strategies, which are integrative strategies designed to provide long-term competitive advantages to organizations by simultaneously enhancing the three dimensions of sustainability. SSM strategies are multilevel and multidimensional, designed to create opportunities for firms by attending to the ecological and social dimensions of their products, services, and processes from cradle to cradle.[1]

A popular framework used by many organizations, such as Shell, Ford, and KPMG, to capture their strategic commitment to all three dimensions of sustainability is the *triple bottom line*.[2] Firms pursuing triple-bottom-line strategies are committed to economic success that both enhances and is enhanced by their concerns for the greater social and ecological contexts in which they exist. Because of organizations' familiarity with the concept of the bottom line, the triple-bottom-line image can be more easily integrated into current organizational cultures. Although circumstances may at times require firms to emphasize one triple-bottom-line dimension over the others, firms basing their long-term strategic direction on the triple bottom line want, if at all possible, to focus their strategic initiatives at the intersection of economic success, social responsibility, and ecological health (see Figure 3.1). This intersection of economy, society, and nature is the playing field of sustainable strategic management and the place where the roles and responsibilities of strategic managers pursuing sustainable strategic management processes and strategies are defined.

CEO Roles and Responsibilities in Sustainable Strategic Management

Chief executive officers (CEOs) are at the pinnacle of the strategic manager hierarchy. Traditionally, CEOs have broad responsibility for determining the long-term direction of the organization, and they are the managers ultimately held responsible for the economic success of the firm. CEOs guide the development of the internal capabilities of the firm in order to make goal achievement possible, and they also manage the strategic processes required to advance their firms toward their visions. The job performance of CEOs is typically measured in terms of their organizations' short-run economic success, even though a growing body of research demonstrates that short term performance measures such as profits and return on assets are poor indicators of future performance.[3] The current business landscape is littered with those CEOs whose job performance was found lacking when measured in terms of the market value of their firm's stock. Of course, introducing sustainable strategic management processes into firms adds social and ecological dimensions to the traditional bottom-line economic roles and responsibilities of CEOs. In this section we examine the roles and responsibilities of CEOs as they relate to the formulation, implementation, and evaluation of sustainable strategic management.

Leading

In today's world of rapidly changing, knowledge-driven organizations, there is little room for autocratic CEOs who bark orders and expect compliance. When Henry Ford began Ford Motor Company, he knew more about building cars than anyone in the firm, so it was appropriate that he managed the company from the position of autocratic expert. However, it is unlikely that Bill Ford knows how to change his car's oil, much less understands all of the technical intricacies and dynamics of modern automotive design, development, production, marketing, and so forth. His role as CEO is one not of

Figure 3.1 **The Triple Bottom Line**

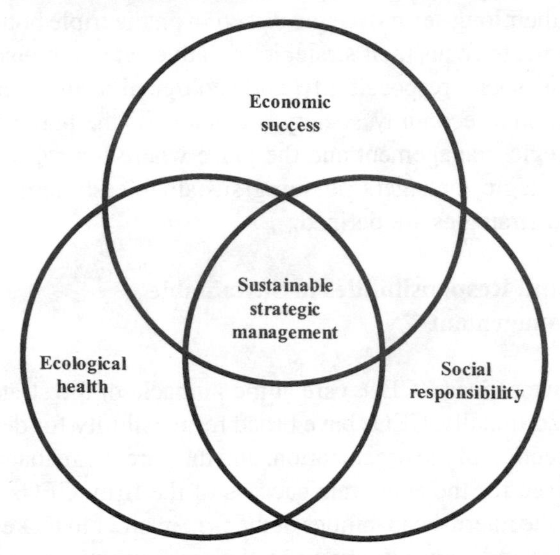

giving orders but of leading Ford as it builds, changes, and faces new complexities, new capabilities, new markets, new technologies, and so forth.

Thus, a central role of CEOs in modern organizations is leadership. By "central" we mean that fulfilling the leadership role is critical to fulfilling the other CEO roles and responsibilities with regard to sustainable strategic management. Since the 1982 publication of the classic *In Search of Excellence*, by Tom Peters and Robert Waterman, a plethora of authors have extolled how CEOs should carry out their role as leaders. In *The Fifth Discipline*, Peter Senge provides a very thorough discussion of the leadership role, saying that effective leaders in modern organizations must serve as designers, stewards, and teachers.[4]

As William McDonough constantly reminds people in his writings and presentations, "Everything starts with design."[5] To design means to mentally and visually conceive an idea, to have a goal or purpose, to make and execute plans. Thus, via their designer role, CEOs reflect their most basic intentions, aspirations, and hopes. Senge points out that design work is seldom visible, is performed behind the scenes, focuses on the future, and involves integrating organizational visions, values, policies, strategies, structures, and systems. Collins and Porras maintain that leaders should be "clock builders" who focus on building the company and what it stands for rather than "time-tellers" who view the company merely as a vehicle for a product-market hit.

Stewards are people who keep something in trust for others. The key to understanding the leader's stewardship role is to recognize that stewards commit themselves to the long-term survival of something larger and more important than themselves. Stewards do not use, waste, and discard for short-term gain; rather they nurture, preserve, and save for long-term survival and success. As such, stewards are servant leaders who serve from the bottom rather than ruling from the top. Their role is to stand in service of the organizational vision, values, stakeholders, employees, and so forth. Jim Collins summarizes the stewardship role of CEOs very well in *Good to Great*, where he states that CEOs of great companies demonstrate humility, modesty, and a personal will, as well as channel their energies into the organization and its shared vision rather than serving their own self-interest. Stewardship seems an especially important leadership function today given the current public distrust created by corporate executives such as Bernie Ebbers of WorldCom, John Riggas of Adelphia, Ken Lay of Enron, Dennis Kozlowski of Tyco, and Sam Waksal of Inclone, all of whom have managed to sacrifice the interests of their corporations for their own personal gain.

The purpose of a teacher is to facilitate learning, and learning is at the heart of what modern organizations do and how they succeed. As teachers, CEOs guide organizational members through processes that will allow them to clearly define current reality, examine current organizational practices, processes, values, and assumptions, and make changes in these organizational elements so that they better match current reality. Among the key responsibilities of teachers are creating structures and processes in which learning can take place and providing for a continuous open, accurate, honest dialogue among organizational members.

Instilling Core Values

As we will discuss in depth in Chapter 7, values are critical factors in strategic decision making. Values serve both as data filters, determining what information strategic managers attend to and how they interpret it, and as decision shapers, providing the frameworks upon which final decisions are based. Values are generally arranged in value systems, which are complex networks of related values. At the center of these networks are *core values*, values that define the essence of the system. Core values are considered good in and of themselves, are typically few in number, and are enduring. Core values provide the overarching ideals upon which organizational value systems are based. Thus, at the heart of the roles and responsibilities of CEOs is the instilling of core values into their firms. These values serve as foundations for the visions and strategies that will determine what organizations will become in the future. For example, Truett Cathy, founder and owner of Chick-fil-A, says that he is on earth to serve, and not just chicken sandwiches. He manages his business according to what he calls biblical principles, where God is his core value. He has built his fortune by putting people before profits, resulting in more than $1.2 billion in sales in 2001.

Recall from our discussion of E.F. Schumacher in Chapter 2 that using linear logic to solve divergent problems leads to seemingly irreconcilable dichotomous dilemmas. Schumacher suggested that solving divergent problems requires identifying overarching, higher-level values that can transcend these dilemmas. Numerous dichotomous dilemmas are posed when the relationships among the economy, society, and ecology are viewed linearly. Growth versus no growth, wealth versus beauty, wealth versus poverty, jobs versus the environment, economic development versus local culture, people as means versus people as ends, and nature as means versus nature as ends are but a few of these dilemmas. Linear arguments like these create the perception that there are inherent, unsolvable conflicts among economic success, social responsibility, and ecological health. Thus, a key role and responsibility of CEOs in firms pursuing sustainable strategic management is to design their firms' value systems around a core value of sustainability, which will allow them to transcend the economy–society–ecology dichotomies and find positive synergy among these dimensions.

Instilling core values requires identifying a set of *instrumental values* that support it. Instrumental values generally define the behaviors and means for implementing and living up to core values. Unlike core values, instrumental values are typically varied and numerous, and they can be more readily questioned and changed than core values can. In Chapter 7, we will present and discuss at length several instrumental values that support a core value of sustainability, including wholeness, posterity, diversity, quality, community, dialogue, appropriate scale, and spiritual fulfillment. This is not an exhaustive list, of course, but regardless of the specific instrumental values, it is important to understand that identifying and internalizing a set of supportive instrumental values is critical to instilling sustainability as a core value in organizations.

Instilling Visions

One critical role of CEOs is to lead their organizations through a vision-building process. Visions are images that portray what firms are, what they stand for, and what they desire to be in the future. Visions serve as foundations for developing missions, goals, objectives, and strategies, and they serve as a basis for determining what information organizations consider important and how they measure success. Thus, strategic visions provide molds through which strategic actions can begin to take shape.

Fulfilling the visioning role requires leading organizations as they build shared visions. This begins with a discovery process that determines what the organization stands for, how it views itself (its self-identify), what factors hold it together, and so forth. Key to an effective discovery process is the creation of a clear image of the firm's current reality, including any hard truths that must be faced. Next comes crafting the vision. True visions (as opposed to slogans) emerge from creative processes that combine an organization's core values and core purposes (its core ideology) into an image of the future complete with challenging goals that require thinking beyond current capabilities (a stretch) and a vivid description of what it would be like to achieve these goals (a dream). This is a process similar to painting a picture with words based on passion, emotion, and conviction. Fulfilling the visioning role allows CEOs to manage the synergy between stability and change by focusing their leadership efforts on preserving their firms' core ideology while at the same time encouraging continuous examination and changes in organizational practices, systems, processes, goals, strategies, and so forth.[6]

An excellent framework for crafting shared visions based on a core value of sustainability is *enterprise strategy*. Enterprise strategy is an overarching strategy that explicitly articulates the firm's ethical relationship with its stakeholders. As such, enterprise strategy applies firms' value systems to the development of corporate strategies, competitive strategies, and functional strategies, allowing firms to base their strategic decisions on their responsibilities to the larger society of which they are a part. Thus, a firm's enterprise strategy is designed to answer the question, "What do we stand for?" The true strength of enterprise strategy is that it explicitly addresses the value systems of managers and stakeholders in concrete terms, focusing attention on what the firm *should* do. Understanding a firm's enterprise strategy requires analyzing three interacting components. The first component is values analysis, designed to uncover the core and instrumental values that compose the firm's ethical system. The second component is issues analysis, which allows the firm to develop a clearer understanding of its social and environmental context. The third component is stakeholder analysis, which helps the firm to identify its various stakeholders and their issues in order to understand what their stakes are and what powers they have to influence the firm.[7]

From an enterprise strategy perspective, CEOs leading their firms toward sustainable strategic management visions facilitate the development of organizational value systems based on a core value of sustainability, providing them with an understanding of the social and environmental issues related to their firms' activities and allowing them to account for the social and environmental concerns of their stakeholders. Such an understanding allows CEOs to guide their organizations toward shared visions that communicate to employees, external stakeholders, and society that their organizations "stand for sustainability" (we present a thorough discussion of "standing for sustainability" in Chapter 7).[8] Such a vision provides the foundation for making decisions that support a new definition of long-term organizational prosperity, one that integrates the need to earn a profit with responsibility to foster social welfare and environmental protection. Like a sailing craft in a long relay race, firms can see that the key to being successful is the efficient and effective use of renewable resources and energy, the ability to be light and maneuverable, the ability to contribute positively to the common good, and the ability to leave no trace of operations in their wakes.

Managing Stakeholders

Another critical role and responsibility for CEOs is managing their firms' stakeholders. Stakeholders include those customers, shareholders, employees, regulators, suppliers, competitors, advocacy groups, and so on that have an interest in the practices of the corporation (we discuss these and other stakeholders and their relationships to sustainable strategic management in Chapter 5). Research has shown that stakeholders either moderate or directly influence the financial performance of firms, that it is possible to align the conflicting interests of multiple stakeholders, and that the importance assigned to particular stakeholders is related to the CEOs' perceptions regarding stakeholder power, legitimacy, and urgency. Managing stakeholders involves serving and balancing their varied and often conflicting needs and interests. One of the primary themes of stakeholder theory is that practicing stakeholder management requires firms to focus on the underlying value systems upon which all of their strategic management processes and decisions ultimately rest. Thus, engaging stakeholders helps CEOs to ground their decisions and actions in the moral and philosophical foundations of their organizations.[9]

The number and diversity of stakeholders that CEOs must manage have increased significantly over the years. Effectively managing stakeholders today requires that CEOs engage in both systems thinking and inclusive leadership. Stakeholder management entails determining who the stakeholders are, how they are changing, what their power and stakes in the organization are, what their expectations and interests are, and what their values are. Strategic managers must analyze the impact that diverse stakeholder groups have on their organization's strategies, and they must develop effective processes that they can use to carry out their organization's purpose with regard to each stakeholder group. Proactive stakeholder management requires the allocation of resources and the delegation of responsibility and authority to those strategic business unit managers who directly interact with various stakeholder groups. Further, in today's turbulent environment, CEOs must develop network management skills in order to effectively deal with the partnerships and alliances they form with suppliers, customers, competitors, and third parties. If strategic managers carefully listen to their stakeholders and view them as sources of information and resources rather than problems to be managed, they can enhance the capabilities of their firms to embrace and manage change because the stakeholders themselves are important agents of change. Thus, CEOs who demonstrate high levels of stakeholder management capability are helping to ensure the survival and success of their firms into the future.[10]

Sustainable strategic management requires reframing stakeholder theory to include the earth as the ultimate stakeholder of business organizations, and it requires recognizing and managing the business environment's stakeholder network, which represents the earth's social and ecological systems (to be discussed in detail in Chapter 5). This expanded perspective on organizational stakeholders provides strategic managers with a framework for discovering and taking advantage of their triple-bottom-line opportunities.

Creating Cultures of Change

Another key role and responsibility for CEOs is to lead their organizations through times of environmental turbulence and change. Strategic managers in the United States often underestimate the amount of time that it takes to implement change, and they often have difficulty practicing the very change processes they advocate. To avoid such issues, CEOs need to ensure that everyone involved understands why change is necessary, and they need to align employees' goals with the goals of the change. They also need to inspire and motivate employees to be a part of the process and to sustain its momentum by empowering them and by evaluating and communicating to them the results of their efforts.[11]

As we discussed in Chapter 1, CEOs who lead sustainable strategic management efforts must serve as change agents, guiding their firms through transformational processes in order to fundamentally shift their organizational cultures to embrace the underlying assumptions and values of sustainability. One important key to creating organizational cultures that support such transformational change processes is to create learning structures that facilitate continuous organizational examination and renewal via personal and shared visioning processes, long-term systems thinking processes, and dialogue-based team learning processes.[12] (We discuss learning structures in Chapter 9.)

Serving as Spiritual Leaders

CEOs guiding their organizations through transformational change processes must serve as both spiritual and strategic leaders. During the early part of the twentieth century, psychologists developed tests for measuring rational intelligence, known as intelligence quotient (IQ). According to the theory, a high IQ reflects a high ability to solve logical problems. During the mid-1990s, it was proposed that emotional intelligence (EQ) is as important as IQ.[13] EQ is a measure of people's awareness of other people's feelings as well as their own. As such, it is the source of human compassion, empathy, and motivation.

There are now indications that there is a third type of intelligence known as spiritual intelligence (SQ).[14] This is the intelligence that humans use to solve problems of value and meaning. SQ helps put their behaviors and lives within a larger context of meaning, and thus it serves as the foundation of both IQ and EQ. Unlike other species, human beings search for meaning and value in what they do because they are driven by questions regarding why they exist and what their lives mean. Humans have a longing to feel that they are part of a larger purpose, something toward which they can aspire. SQ allows them to be creative, to use their imaginations, and to change their rules. It allows them to think out of the box and to play with the boundaries of their existence. It is this transformative characteristic that distinguishes SQ from IQ and EQ. Whereas both IQ and EQ work within the boundaries of the situation, SQ allows individuals to question whether or not they want to be in the situation in the first place. SQ facilitates the dialogue between reason and emotion, between mind and body. It provides the ability to integrate all the types of intelligence. Thus, it is a transcendent intelligence.

Effective sustainable strategic management requires all three types of intelligence. IQ is needed to facilitate the strategic and logical reasoning necessary for maintaining the economic success of the firm. EQ provides a foundation for effective stakeholder management, caring for others who have a stake in the organization. In addition, EQ enables CEOs to empathize with the people of the world who are affected by the inequitable consumption of resources and distribution of wealth. But it is SQ that visionary leaders need to inspire others to feel that they are part of the larger purpose of sustainability. Recall that both Aldo Leopold and E.F. Schumacher (see Chapter 2) maintained that the transformation to a sustainable world would require humans to transcend to a higher, more spiritual level of consciousness regarding their relationships with one another and with nature. It is SQ that provides CEOs with this higher spiritual perspective. Moreover, the transcendent nature of SQ allows CEOs to integrate all three types of intelligence, providing them with a framework to look beyond the boundaries of the present and ask what can be (see Figure 3.2). As such, CEOs with high levels of SQ can inspire others to think outside of their existing boundaries, and to envision a sustainable firm in a sustainable world.

Roles and Responsibilities of Boards in Sustainable Strategic Management

Traditionally, boards of directors have had the fiduciary responsibility to represent the owners of the company and to ensure the long-term economic success of the firm. This entails overseeing the strategic management processes of formulation and implementation of the firm's vision, mission, goals, and strategies. In sustainable strategic management the responsibilities of boards expand to include social and ecological as well as economic performance of the firm. As is the case for CEOs, this will require boards to probe and question the fundamental assumptions underlying their firms' strategic initiatives.

Board members have the potential to add unique external perspectives as firms attempt to push their existing boundaries and create new triple-bottom-line opportunities. That is why John Elkington says that boards are the best vehicles for incorporating sustainability into core assumptions and values of firms (what he calls firms' DNA).[15] However, getting and using these unique perspectives from boards is not automatic. To provide these unique perspectives, board members of organizations pursuing sustainable strategic management should have broad backgrounds that collectively represent the economic, social, and ecological interests of the firm, meaning that firms will have to go outside the shareholder ranks and add board members who represent the interests of social and ecological stakeholders. Fortunately, adding such nonshareholder board members is a growing trend that has already become institutionalized in large public corporations.[16] The board members chosen to represent society and nature should be both prepared and allowed to voice their perspectives and have them incorporated into the strategic choices of the firm. This means that these board members will have to take active learning roles rather than rubber-stamp roles in the organizations they serve. These learning boards will have to keep up with the changes in the turbulent external environment, they will have to focus continuously on the triple bottom line, and they will have to work closely with CEOs to find paths to triple-bottom-line success.

Figure 3.2 **Sustainable Strategic Management Intelligence**

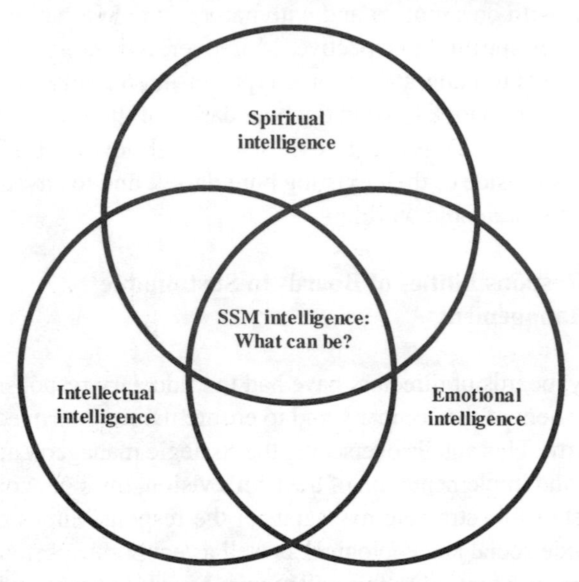

In addition to overseeing the long-term strategic direction of firms, boards are also responsible for the ethical and legal ramifications of their firms' activities. The recent corporate scandals in the United States have made abundantly apparent how important this board responsibility is. These scandals created a crisis of trust and led to the passage of the Sarbanes-Oxley Act in 2002, legislation that requires financial integrity and accuracy in public companies. These scandals and legislation have forced many boards to do some hard soul searching about the values of their organizations. There are many legal and ethical dimensions related to sustainable strategic management, and there are many questions that boards must ask in order to fully understand these dimensions. Boards will want to explore questions related to the legal requirements and fiduciary responsibilities of their firms related to environmental and social performance (pollution, civil rights, product safety, etc.). They will also want to explore several issues related to where they stand ethically. For example, they will want to explore their level of commitment to future generations, and they will want to know how they can balance the pressures for short-term profits with the long-run demands of sustainability. Of course, these are questions of value and meaning. Thus, like CEOs, boards of directors overseeing organizations pursuing sustainable strategic management require a high level of spiritual intelligence to function effectively.

Boards are also responsible for auditing the performance of their firms and CEOs; this is essentially their watchdog role. The selection and succession of CEOs as well as the determination of CEO pay also lie in the hands of the board. Recent business scandals in the United States unveiled an interlocking web of directors that demonstrates a high concentration of power in the hands of a few. Eleven of the fifteen largest companies have at least two board members who sit together on another board. Twenty percent of the 1,000 largest firms in the United States share at least 1 board member. More than 1,000 people sit on 4 or more corporate boards, and 235 sit on more than 6. For example, Sandy Weill, former CEO of Citigroup, and Michael Armstrong, former CEO of AT&T, sat on each other's boards. This CEO interlock raised many ethical questions. Weill admitted that in 1999 he had asked Salomon Smith Barney's star analyst, Jack Grubman, who had a negative rating on AT&T stock, to take another look at it. Grubman ultimately upgraded his rating to a strong buy, thus influencing Armstrong to support Weill in a power struggle at Citigroup.[17] Other effects of interlocking boards have been demonstrated in many cases of exorbitant CEO pay (often not in line with firm performance) and in the abuse of stock options among top managers. With so much power resting in the hands of a few board members, there are concerns that the audit function of boards can be compromised, especially insofar as so many CEOs in the United States also serve as chairpersons of their boards. Objective audits of firms' performance in terms of economic, social, and ecological criteria are critical for successful sustainable strategic management. The better the system of corporate governance is, the greater the chance that firms can move toward genuine sustainability.[18]

Accountability and transparency of the board of directors has become increasingly important. Sustainable strategic management requires that boards take a holistic perspective with respect to a broad range of accountability issues. The board's fiduciary responsibility to shareholders has expanded to include a dialogue with many other stakeholder groups. Many board members fear that they will be held personally liable, risking their own assets or being imprisoned for the actions of the corporations they serve. Board members can be held legally accountable for developing nonrecyclable packaging in Germany, cutting hardwoods in Canada, importing ivory from Asia or East Africa, or polluting waterways in England.[19] Texaco's record-setting race discrimination settlement of $176 million was a wake-up call to many boards that they can be held accountable for their firm's social performance as well as its economic and environmental performance.

Missions, Goals, and Strategy Formulation in Sustainable Strategic Management

After building a shared vision around a core value of sustainability, it is the responsibility of strategic managers to define the organization's mission and goals and to formulate SSM strategies to fulfill the mission and meet the goals. As discussed earlier, the vision of sustainability is reflected throughout the strategy formulation process. This is particularly important in mission definition because the mission defines the organization's unique purpose. It addresses the question of why the firm exists. The mission statement sets the organization apart from others of its type and identifies the scope of the firm's operations, usually in product and market terms. In other words, the mission statement defines the organization's business. At the competitive level the mission focuses on customer groups and needs, while at the corporate level it focuses on the purpose, scope, and balance of the firm's portfolio of strategic business units.

Many firms today align their business definition to a noble purpose in an attempt to inspire pride among their stakeholders.[20] For example, Merck Pharmaceuticals defines its business as "Preserving and improving human life," British Petroleum's new brand identity is "Beyond Petroleum," and Mary Kay Cosmetics' mission is "To enrich women's lives." A core value of sustainability can certainly inspire a noble purpose that encourages employees as well as other stakeholders to think openly about the meaning of the firm's work and its connection to the greater social system and ecosystem. Many firms today embrace the concept of sustainability and reflect it in defining their businesses. For example, Shell notes in its 2000 report, "We wish to play our part responsibly by maintaining and enhancing natural and social capital, as well as contributing to the global economy's capacity to generate and distribute wealth." BASF has also issued a statement in which it pledges to use the principles of sustainability in its strategy formulation. AOL Time Warner says in its 2002 *Social Responsibility Report* that it is committed to the principles of sustainability in all of its business units in order to serve and strengthen society.[21] The fact is, many firms are jumping on the sustainability bandwagon, using sustainability as a framework for their reports to stakeholders. Of course, the real test of this commitment is how well strategic managers in these firms "walk their talk" and actually measure their success in terms of social and ecological as well as economic performance.

After sustainability is integrated into the vision and mission of the firm, goals are formulated that specify the long-run results that the organization seeks to achieve in terms of sustainability. These provide a broad, general direction for the firm within the context of the vision and mission. Because of the abstract nature of sustainability, the formulation of specific goals is critical to the success of sustainable strategic management. Only by translating sustainability into specific performance targets can strategic managers begin to actively manage in a sustainable manner. Goals provide an index for measuring the firm's progress with respect to its social, ecological, and economic performance, so it is imperative that the organization have explicit goals in each of these areas. For example, Nike has published its short-run and long-run goals for environmental sustainability. Some of its short-run goals include: decreasing throughput of one-way resources, designing products that can either be donated, bought back, remanufactured, or returned to nature, and changing the emphasis from human productivity to resource productivity. Some of Nike's long-run goals (for 2020) include: closing the loop and taking full responsibility for their products, eliminating all substances that are known to be harmful to human health or the health of biological or ecological systems, and developing financial structures that promote greater product stewardship. As Sarah Severn, director of sustainable development for Nike, says, "If we accomplish all this, and that's a big if, we will have a business that is prosperous in the long term, and we will have acknowledged the needs of future generations."[22]

Crafting SSM strategies requires a transformation in the planning process. SSM strategies are plans that are formulated in order that the firm can achieve its sustainability goals (we will expand on the content of these strategies in Chapter 6). The processes that are utilized in formulating SSM strategies require that strategic managers encourage systems thinking throughout their organizations. The planning processes of the past were seen as activities to be performed at an annual planning retreat and were typically expert- and staff-driven. The planning goal was the creation of a strategic plan. As the turbulence increased in the business environment, strategic managers realized that planning needed to be an adaptive, continuous process of matching external opportunities and threats with the firm's strengths and weaknesses. In the current process models, managers from the top to the line take an active role in the strategy formulation process. The planning goal is usually the incremental improvement of market share within existing industry boundaries with a focus on "what is," reflecting an adaptive organizational learning process.

Current models of strategy formulation assume that the firm operates within an economy that is closed to the greater society and ecosystem. Crafting SSM strategies requires a shift in the mental model of strategic managers from this closed circular flow view of the economy to an open living system view (as discussed in Chapter 1). This new paradigm enables strategic managers to structure their firms' strategy formulation processes around the core value of sustainability. An open systems model of planning requires the involvement of many managers, suppliers, customers, and other important stakeholders who represent the collective wisdom of the firm. Such a process facilitates the transition from adaptive to generative organizational learning, allowing for the continuous questioning of fundamental assumptions and values related to why and how firms are doing business. These generative planning processes allow firms pursuing sustainable strategic management to rewrite industry rules and to think beyond existing industry boundaries, enabling them to focus on "what can be." Thus, SSM strategy formulation processes expand organizations' ability to create their own future by questioning the underlying assumptions and values on which strategy formulation is based.

Conclusions

As we come to the end of this first section of the book, we would like to summarize some of its important points. Our purpose in this section was to establish a foundation for sustainable strategic management. In doing so we discussed in Chapter 1 that economic activity on the small planet earth cannot continue to operate under the assumption that the economy is a closed circular flow with an unlimited supply of natural resources, an unlimited supply of cheap labor, and an unlimited capacity to dispose of wastes. We contended that business activity should instead be based on the assumption that the economy is an open living system that can survive in the long run only by bringing economic activity into balance with the needs of society and the limits of nature, and we made the case that strategic managers in business organizations are among the people who can most influence this shift in the economic paradigm.

In Chapter 2, we presented an in-depth discussion of sustainability, the foundation of an open living system economy and the underlying core value for sustainable strategic management. We discussed some of the scientific underpinnings of sustainability, we discussed the economic, social, and ecological dimensions of sustainability, and we presented some of the ethical frameworks supporting sustainability. We concluded the chapter by pointing out that weaving sustainability into the strategic fabric of organizations will require changing not only the way strategic managers do business but also the way they think about business.

In this chapter, we took an in-depth look at the definition of sustainable strategic management, we discussed the roles of CEOs and board members in leading and overseeing sustainable strategic management efforts in organizations, and we discussed the overall responsibilities of strategic managers related to formulating missions, goals, and strategies that support sustainable strategic management. In the next section, we will look more closely at the formulation of sustainable strategic management. We will discuss environmental analysis and strategic advantage analysis as they relate to sustainable strategic management formulation, we will discuss the levels and content of SSM strategies, we will discuss strategic choice processes, and we will provide a value-based framework for making strategic choices that are compatible with sustainability.

Part II

Formulating Sustainable Strategic Management

Now that we have defined sustainable strategic management, introduced some of its key foundational frameworks and concepts, and discussed the key roles and responsibilities of sustainable strategic managers, we want to discuss the formulation of sustainable strategic management strategies (SSM strategies) in some depth. In doing so, we will discuss external environmental analysis as it relates to identifying sustainability-based threats and opportunities, focusing on analytical frameworks, tools, and processes that can help organizations understand and manage the sustainability issues facing them in both the macro and competitive environments. Next, we will discuss strategic advantage analysis, focusing special attention on how value-chain analysis and stakeholder analysis can be used as means for identifying an organization's strategic advantages. Then we will provide an in-depth examination of the content of SSM strategies at the functional, competitive, and corporate levels, and we will follow that with an in-depth look at the complex value-laden cognitive processes related to making sustainable strategic choices. We will conclude this section with an integrative model of the sustainable strategic management process.

4

Environmental Analysis for Sustainable Strategic Management

E.F. Schumacher described the Arab oil embargo of 1973 as the watershed economic event of the twentieth century, marking the beginning of an unprecedented era of turbulence in business that continues today. Schumacher said, "Things will never be the same again,"[1] and they have not. The embargo provided the small crack that many organizations (e.g., Japanese auto manufacturers) needed to get a stronger foothold in the international market. From that point the global economic race was on. A key issue exposed by the embargo was that the causes and effects of environmental turbulence go beyond the boundaries of economic activity, spilling into the arenas of politics, social welfare, and ecological concerns. Incredible advances in technology have paralleled the explosion of global economic activity since the embargo. Modern information technology has transformed almost every product and every process in every industry. Furthermore, there have been increasing social and environmental demands from citizens worldwide, leading to unprecedented levels of consumer advocacy, social activism, and legislation aimed at changing the way organizations do business. These complex, interrelated economic, technological, social, and ecological demands define the modern business environment.

Igor Ansoff identified four factors that make today's business environment different from the past.[2] First, the speed of change in today's business environment is very fast. Many use terms such as "permanent whitewater" to describe how fast and turbulent the changes faced by organizations are today. Second, so much of the change happening in the current business environment is novel and discontinuous, meaning that predicting the future based on the past is very difficult, even impossible at times. This renders traditional forecasting techniques less effective for predicting future trends, and it makes it even more important that organizations develop strategic management processes that encourage questioning and changing mental frameworks, allowing them to view things differently as well as to do things differently. Third, as pointed out above, change today is occurring on a global scale. And fourth, the number and complexity of critical success factors is increasing significantly. Today strategic managers must focus not only on putting the right products and services in the right markets but also on myriad nonmarket factors such as human rights, complex regulatory frameworks, technological change, resource constraints, environmental responsibility, and so forth.

Given the complexity of today's business environment, it is imperative for strategic managers to develop environmental scanning cultures within their organizations. Determining environmental opportunities and threats should result from the collective wisdom of the firm's stakeholders. Successful sustainable strategic management requires that opportunities and threats be identified, and it requires that they be analyzed in terms of the underlying assumptions on which they are based. Although most strategic management texts typically segment the external environment into two sectors (macro and industry) for analysis purposes, this does not mean that these environmental forces are mutually exclusive. The fact is, the forces in the environment are so dynamic and interrelated that environmental analysis must be based on systems thinking, focusing on developing information flow, feedback loops, analytical processes, dialogue processes, and so forth that will allow organizations to recognize, understand, and capitalize on the environmental turbulence that surrounds them.

Macroenvironmental Analysis

The macroenvironment is sometimes referred to as the remote or general environment because it affects virtually all organizations, and strategic managers have little or no influence over it. The macroenvironmental forces typically covered in strategic management texts include economic, technological, sociocultural, demographic, and political/legal. The convergence of these forces in the marketplace provides many opportunities as well as threats for business organizations. For example, several opportunities, threats, and strategic issues arise from the convergence of the economic interdependence of global economic markets, the desire of firms to increase their scope of operations, and the social and economic dislocations from shifts in the demographic sector. Eighty-five percent of the world's population will live in developing countries by 2025.[3] The average age of the population in these countries is very young, so the working-age population is expected to grow rapidly. This represents major opportunities for business with respect to new labor and consumer markets. Because of the low birth rates and high longevity in the industrialized countries, however, there will be a prominence of people aged fifty to ninety years old. These population trends have a tremendous impact on health care, social services, pensions, and so forth in the industrialized world, where a smaller population of young people must support these services. In order for strategic managers to take advantage of the new labor and consumer markets in the developing countries, they will have to add health care, education (especially of women), family planning, water infrastructure, and so forth to their strategic agendas. Only if the population is healthy and educated can business take advantage of the opportunities that these new markets offer.

Another example is the threats and opportunities that arise from the convergence of the technological revolution with socioeconomic forces. Currently, more than half of the world's population has never used a telephone, only 7 percent have access to a personal computer, and only 4 percent have access to the Internet. These macroenvironmental trends provide real opportunities for business to offer services to billions of people, allowing them to participate in the world economy by connecting them with distant people and markets. However, Internet users must be literate, and in today's world one out of every five adults is functionally illiterate.

Opportunities and threats also arise from the convergence of the interdependence of the global economic markets with geopolitical forces and trends. While the disruption of the global socioeconomic system from war and terrorism creates many global threats, the spread of democracy creates many opportunities in the market-based economies. The number of democratic states has increased from 22 in 1950 to 119 in 2000, and the proportion of the world's population living in freedom has increased from 36 percent in 1981 to 41 percent in 2000. This means more secure conditions for business operations, growth, and investment.

The Issue Wheel: Including Nature in the Macroenvironment

As demonstrated in the above examples, there are numerous opportunities and threats facing business organizations resulting from the convergence of the traditional macroenvironmental forces covered in strategic management books. However, most of these texts are implicitly based on the assumption that the economy exists in an unlimited, bountiful, ever-resilient ecosystem. Strategic management scholars have long recognized society as an important macroenvironmental force, but most have omitted the geographical location in which all business takes place, the natural environment.[4] Because humankind, and thus business, is highly interconnected with nature, inclusion of the ecological sector in environmental scanning and analysis is critical for the systems-based approach necessary in formulating sustainable strategic management strategies. In this section, we examine nature's role in the business environment.

At the center of the environmental impacts related to humankind's activities are the interactions among three variables: population growth, the level of human affluence as measured by the growth in per capita GDP, and the impact of technology on the natural environment as measured by the materials and energy used and the pollution and degradation generated for each unit of GDP created. Like a wheel, the environmental issues faced today begin with the interactions among these three central variables and radiate out into the larger ecosystem. As depicted in Figure 4.1, population growth, economic growth, and technology are acting together to create rapidly increasing production and consumption. As humankind produces and consumes in order to meet the demands of an exponentially growing number of people, it continues to deplete its resources and foul its nest with its own wastes and pollutants. The resulting problems include environmental catastrophe, poor air and water quality, loss of species, climate change, land degradation, deforestation, wetlands loss, human health problems, a lower quality of life for many, and so forth, now and for posterity.

Population, Economy, and Technology at the Hub

Recall from our discussion in Chapter 2 that the global population has multiplied sixfold in the past 150 years and is growing at an annual rate of 1.3 percent—77 million or so new people on the planet each year.[5] An important point to understand when examining population statistics is that overpopulation is not really a problem of space. At the heart of the problem are the vast quantities of natural resources used and wastes generated by the human species as a result of its activities, often referred to as the human "footprint." Despite the fact that most population growth is occurring in developing nations, when resource use and waste generation per person are taken into account, overpopulation becomes a serious issue in the developed world as well. For example, it is often reported that each child born in the United States will have up to thirty times the environmental impact as a child born in India. Ehrlich and Ehrlich use the Netherlands as an excellent example of this. They point out that the Netherlands—a densely populated developed nation—imports massive tonnages of various foodstuffs and 100 percent of its iron, tin, bauxite, copper, and many other minerals; furthermore, it extracts most of its fresh water from the Rhine River, which flows from

Figure 4.1 **The Issue Wheel**

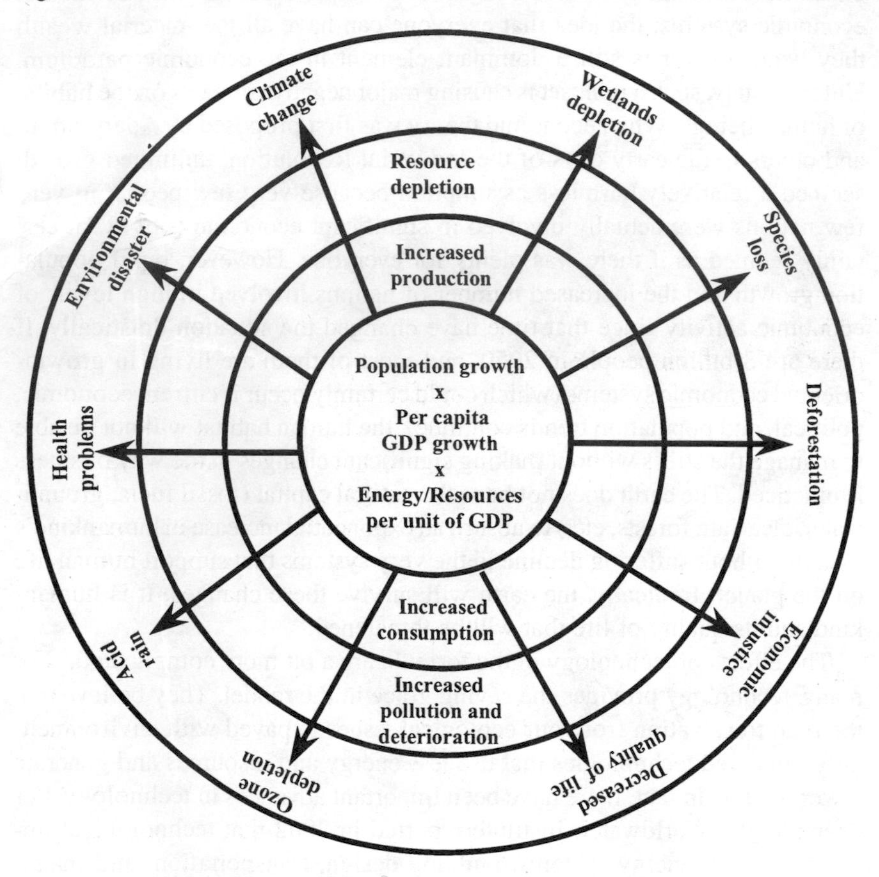

Source: Adapted from W.E. Stead and J.G. Stead, *Management for a Small Planet* (Thousand Oaks, CA: Sage, 1996).

countries to the north. Until productive natural gas fields were discovered in the 1970s, the Netherlands also imported all of its energy; these gas fields will be exhausted soon, and the Netherlands will again be 100 percent dependent on energy imports. Thus, the Netherlands can support its dense population only by extracting resources from other parts of the world. Overpopulation exists when a region cannot sustain itself without rapidly depleting its resources, degrading the environment, or importing energy and resources from elsewhere to support the population.

Also recall from Chapter 2 that per capita GDP has risen threefold since 1950, and there has been a sevenfold increase in world trade since 1970. The fact is, such statistics are generally reported with glee. The potential for

unlimited economic growth is a key belief that pervades current free market economic systems; the idea that everyone can have all the material wealth they want forever is still a dominant element in the economic paradigm. Unfortunately, such a mindset is causing major negative impacts on the habitat of human beings. When economic theory was first proposed by Adam Smith and others in the early days of the Industrial Revolution, unlimited growth seemed a relatively harmless assumption because very few people in very few nations were actually involved in significant economic activity; it certainly seemed as if there was plenty for everyone. However, rapid population growth and the increased number of nations involved in high levels of economic activity since that time have changed the situation drastically. If there are 8 billion people in 2050, and most of them are living in growth-oriented economic systems (which could certainly occur if current economic, political, and population trends continue), the human habitat will not be able to manage the stress without making significant changes in the way business is practiced. The earth does not have the natural capital (fossil fuels, groundwater, clean air, forests, etc.) to absorb an exponential increase in humankind's assault without suffering decline in the very systems that support human life on the planet. Ironically, the earth will survive these changes. It is humankind and its quality of life that will be threatened.

The effects of technology in the formula are a bit more complicated.[6] For many, technology provides the saving grace in the model. They believe that the road to salvation from our ecological issues is paved with environmentally improved technologies that use less energy and resources and generate fewer wastes. In fact, there have been important advances in technology. For example, the Worldwatch Institute reported in 2002 that technological improvements in energy systems, building design, transportation, and manufacturing have served to reduce greenhouse emissions at a faster rate than that anticipated at the Rio Summit in 1995. However, skepticism abounds with regard to the potential for technology alone to provide the way out of humankind's ecological ills. Statistics show a continued rise in many toxins and pollutants, a continued loss of many species, and a continued decline in many critical natural resources. We will examine these issues in some detail in the next section.

Resource Depletion and Pollution Radiating from the Hub

We discussed previously the fact that the earth is a relatively closed system with little ability to import inputs or export outputs beyond its boundaries. Thus, the increased production and consumption associated with a growing population in a growth-oriented society flies directly in the face of ecological

balance. Resources are being depleted at increasingly rapid rates, and the environment is being degraded by an overflow of pollutants and waste products.

Depleting Resources. Over its 4.5 billion years of existence, the planet has assembled an abundance of natural resources, such as fossil fuels, high-grade mineral ores, trees, rich agricultural soils, groundwater stored during the ice ages, and millions of species. Resources are critical to humankind's survival because, ultimately, they are the only capital humans have. They are the basic source of material wealth, financial wealth, and psychological wealth. The news on how well humans are protecting their natural capital is not all that encouraging.[7] In Chapter 2, we mentioned the potential for severe water and cropland shortages in the coming years. In fact, humans are violating a basic financial principle familiar to any investor or business student: One must live off income, not capital; living off capital depletes one's potential for future income generation, and ultimately is a sure route to poverty. Today humans are living off their natural capital, and the results are predictable—rapid depletion of resources and the real potential for ecological poverty.

In addition to water and cropland, the planet's forests continue to disappear at an astonishing rate—a long-standing trend. About 4.2 percent of the world's natural forests disappeared during the 1990s—161 million hectares. It is even more disturbing that 152 million of these hectares were lost in the world's tropical rainforests—once referred to by Brazilian environmental martyr Chico Mendes as "the lungs of the planet" because they absorb so much carbon dioxide and produce so much oxygen. Despite such heavy losses, only 36 million of these lost hectares were replaced with regrowth, and only 10 million of those regrowth hectares were in the tropics. Even more startling is that 89 million of the hectares of natural forest cover lost during the 1990s occurred in only eight countries, Brazil, Indonesia, China, Zambia, Sudan, Mexico, the Democratic Republic of Congo, and Myanmar (formerly Burma). With these forests go the habitats of most of the earth's species, the water and watersheds for billions of people, an important line of defense against climate change, and so forth. Economic activities are the primary culprit in this problem. The forests are being harvested and shipped to developed nations where they are converted into wood for furniture, homes, and so on. Forest clearing is also occurring in order to convert the land for farming and raising livestock. The economic life span of a cleared tropical forest is only a few years because once the tree cover is lost, the soil falls victim to the overwhelming tropical heat. Thus, farming and grazing activities can be supported for only a short time before the sun renders it parched and useless.

Wetlands are another land resource that is rapidly disappearing. Wetlands are vital ecosystems that help to stem floods and erosion, replenish groundwater aquifers, and facilitate the settling or removal of inorganic matter and organic microbes. Furthermore, wetlands, like tropical forests, are rich in biodiversity. Half of the world's coastal and inland wetlands have been lost over the past century, falling victim to construction, channeling, draining, and pollution. For most of the twentieth century, wetlands loss was confined largely to the developed nations of the northern hemisphere. However, this trend has reversed recently, and many fragile wetlands in developing nations in the southern hemisphere are now coming under serious stress because of unchecked economic development.

Nonrenewable fossil fuels are also being exhausted at incredible rates, and this trend is continuing. Since 1950, worldwide coal consumption has doubled, oil consumption has increased 7.5 times, and natural gas consumption has increased over 13 times. These trends dictate the continuous need for new sources of precious fossil fuels, and these sources are invariably less accessible and more costly, both economically and environmentally (e.g., the Alaskan National Wildlife Refuge). In addition, largely because of the concentration of available supplies of these fossil fuels in a few of the world's nations, they are associated with myriad political, social, and environmental issues related to their control, pricing, transportation, and so forth.

Increasing Pollution. So there is little doubt that humankind is overspending its natural capital. But where does it go? The answer, of course, is that it goes into the world's energy, products, and services with its residue often released into the air, water, and ground in the form of environmentally degrading pollution. These are high-entropy human wastes—the stuff emitted into the environment as a result of human's consumption and production processes.

Let us begin by examining air pollution. Fossil fuel burning, biomass burning (i.e., burning the rainforests to clear them for grazing land), agricultural activities, along with declining forest cover are infusing large quantities of toxins into the atmosphere every day. During the Industrial Age, there has been an unprecedented increase in the presence of potentially dangerous trace gases in the atmosphere; trace gases are those that in nature make up a very small proportion of total atmospheric gases. Examples of trace gases that are increasing to potentially dangerous levels in the atmosphere include sulfur and nitrogen compounds, including sulfur dioxide (SO_2), nitrogen oxides (NO and NO_2), and nitrous oxide (N_2O). Others include methane (CH_4) and various chlorofluorocarbons (CFCs).

By far the most prominent trace gases in the atmosphere are from carbon emissions—especially carbon dioxide (CO_2). Carbon emissions have increased rapidly, and these are the emissions most responsible for the current global warming trends (which will be discussed later). The primary responsibility for carbon emissions lies with industrialized nations. At the beginning of the Industrial Revolution, three million tons of carbon were released into the air each year as a result of fossil fuel burning. That figure now stands at approximately 650 billion tons per year. The concentration of carbon in the atmosphere has also risen—increasing from 317 parts per million in 1960 to 370 parts per million in 2000 (a 16 percent increase). There is some silver lining in the carbon emissions story. Carbon intensity—the ratio of carbon emissions to gross world product—has declined from a high of 250 in the 1960s to 150 in 2000 (40 percent). This indicates a trend toward cleaner, more efficient economic processes and technologies. However, in absolute terms, carbon emissions are still at dangerous levels and rising.

Humankind is being no kinder to its water than it is to its air. Besides using water at irreplaceable rates in many corners of the globe (as discussed in Chapter 2), toxic chemicals and other waste products have been poured into the earth's waterways over the years as if they were humankind's personal dump. Human and animal wastes, agricultural wastes, and industrial wastes are too often dumped into the earth's aboveground and underground waterways with little regard for those downstream, and this has led to a major decline in the availability of safe drinking water throughout the world. Of the total amount of wastes generated from farms, homes, and industry, less than 10 percent is treated before it enters rivers and streams used for drinking water and sanitation. Nitrates, pesticides, petrochemicals, arsenic, chlorinated solvents, radioactive wastes, and fluoride are but a few of the substances that find their way into waterways across the globe.

One of the most troubling water issues is the pollution of groundwater in agricultural regions because of increases in the amount of nitrogen applied as fertilizer and huge amounts of animal wastes. Nitrogen use in fertilizer has increased twenty times since the 1950s, and about half of this nitrogen never makes it to the plants. Rather, it is diluted with rain and seeps into the underground aquifers from which the farmers and communities get their drinking water. The wastes from cows, pigs, chickens, horses, and other farm animals far exceed the amount of human wastes generated. In the United States, farm animal wastes are 130 times that of human wastes. Besides running into aboveground waterways, these wastes also dilute during precipitation (especially the nitrates) and seep into groundwater sources.

Many of humankind's wastes must be buried. The images of overflowing landfills and garbage barges cruising the waterways looking for communities willing to take their trash are fresh in the human psyche. In addition to the landfill problem, there is the issue of disposing of hazardous wastes. There are 300 million to 500 million tons of hazardous wastes—wastes that demonstrate one or more hazardous characteristics—generated each year.

Some 80 percent of these wastes are generated in industrial countries, but not all of the wastes are disposed of in those countries. Despite the Basel Convention of 1989, which was designed to stop the shipment of hazardous substances across international borders, 10 percent of the hazardous wastes produced are shipped to other countries for disposal. The countries most actively involved in this practice are the United States, the United Kingdom, the Netherlands, Germany, and Australia.

Although they do not come close to matching the amount of chemical wastes generated, nuclear wastes represent a very serious hazardous waste disposal problem facing humankind. Many of the wastes and by products from nuclear processing facilities, power plants, and so on are dangerously toxic and will remain so for hundreds of thousands of years, and yet no country that produces nuclear wastes has come up with an acceptable solution for their disposal. For example, in the United States a battle has raged for years over the appropriateness of burying nuclear wastes under Yucca Mountain, Nevada, and there really does not seem to be an end in sight. In the meantime, millions of containers of nuclear wastes are stored on the grounds of U.S. nuclear power plants and processing facilities awaiting final disposal.

The Wheel of Misfortune

Thus, as can be seen from the above discussion, humankind is spending its natural resources more rapidly than they can be replenished, and it is polluting its air, water, and land in the process. In short, people are bankrupting themselves of the natural capital on which they depend for survival. As a result, several environmental problems have arisen because of these resource depleting and polluting activities. Let us examine some of these environmental problems, which are portrayed on the outer rim of the issue wheel in Figure 4.1.

On August 7, 1978, President Jimmy Carter declared Love Canal, New York, the United States's first nonnatural disaster. Underneath an elementary school and a residential area with more than 200 homes in the community, a toxic soup of chemicals had risen to the surface. This disaster was the end result of a gaggle of mistakes made by Hooker Chemical Company, the City of Niagara Falls, New York, and many others over a period of thirty-five years. Love Canal quickly became the most famous symbol of environmental disaster in the world. Environmental disasters since then read like a list of nations, towns, and waterways on a perverted guided tour: major oil spills in Monterey, California, Valdez, Alaska, Galveston Bay, Texas, the Shetland Islands, and most recently off the coast of Spain in the prominent coastal fishing region of Galicia; chemical disasters at Times Beach, Missouri, the Sacramento River in California, and Bhopal, India; and nuclear meltdowns at Three Mile Island, Pennsylvania, and Chernobyl in the former Soviet Union.

Even though these environmental disasters become major media events for a time, they are only the tip of the iceberg in terms of serious environmental issues facing the planet. Whereas single spectacular events capture people's attention, insidious long-term problems are often ignored. Of the many insidious environmental problems plaguing the earth today, climate change may be the most serious and is certainly the most immediate.[8] The increasing trace gases in the atmosphere discussed above—especially CO_2—cause solar rays that normally would be reflected from the earth's surface and then released back into space to be trapped in the earth's atmosphere, adding more heat than normal to the atmosphere and leading to rising global temperatures. These gases act like the transparent roof on a greenhouse, hence the term "greenhouse effect." Essentially, the greenhouse effect is occurring because there is no longer a balance in the planetary carbon cycle. In its natural state (pre–Industrial Revolution) the carbon cycle effectively circulates or stores some 42 trillion tons of carbon between the planet's atmosphere, oceans, and land. However, because of the huge rise in carbon emissions from the burning of fossil fuels (see the previous section), more carbon is being trapped in the atmosphere than can be accounted for by the natural carbon cycle. This creates an abundance of greenhouse gases that serves to keep the earth's heat from escaping the atmosphere, resulting in rising temperatures now and into the future. Scientists have found that CO_2 levels are higher now than at any time in the past 420,000 years, and they have shown that CO_2 levels and global temperatures have closely paralleled one another during that period. Moreover, data indicate that global temperatures have only begun to rise. The oceans have warmed significantly over the past forty years, and much of that warming has taken place in the oceans' deepest regions. According to scientists, deep water warming is a clear sign of impending temperature increases, typically preceding a significant rise in temperatures by about ten years. Clearly, this is a problem that humankind has already willed to its children.

There has certainly been plenty of debate over the potential for and the outcomes of climate change. However, the data continue to be compelling, and the risks of not tackling the issue are huge. Will the average temperature rise eight degrees or only one degree? Will it take 50 years or 150 years for the process to have a severe impact on humankind? Is it worth risking natural disasters such as increased hurricanes, dramatic changes in the availability of farmland, the destruction of thousands of ecosystems, and the sinking of vast coastal lands into the sea? Can humankind afford to wait to see for sure? We mentioned briefly in Chapter 2 that precaution is a foundation assumption of sustainability. Global warming is one problem that gives this assumption clear and present meaning. If humankind does not find some way to deal effectively with climate change now, then future generations could suffer greatly from our lack of precaution.

Global warming is but one of many serious environmental problems that have resulted from the interactions of the factors at the hub of the issue wheel. In the past two decades or so, scientists have discovered huge holes in the planet's stratospheric (upper atmosphere) ozone layer, which shields the earth's inhabitants from the debilitating effects of sun's ultraviolet rays, and they have discovered that patterns of acidic precipitation have severely damaged vegetation on some of the planet's most beautiful mountains and in forests, and killed the fish in many of its lakes. Smog, the accumulation of excessive amounts of lower atmosphere ozone, has been tied to human eye and lung diseases, loss of foliage and agricultural crops, the destruction of public landmarks, and other problems. Biodiversity is another critical issue—some argue the most critical because of the remorseless irreversibility of the process. The alarming decline of species in the tropics because of deforestation and poor farming practices is well documented. Birds have been shown to be in danger worldwide, and recent data indicate that 20 percent of the freshwater fish species are endangered because of pollution, dams, and water diversions. Furthermore, there are direct human health costs related to the earth's depletion and pollution problems. A high percentage of the planet's people are exposed to dangerous levels of air pollution, many do not have safe drinking water, and safe sanitation facilities and processes are absent in many parts of the globe. Some of the health costs of conditions such as these include cancer, birth defects, malnutrition, radiation poisoning, respiratory problems, chronic bronchitis, emphysema, dysentery, cholera, and typhoid. Moreover, these health costs are more often borne by laborers, the poor, the underprivileged, and the underserved.

As we draw this section to a close, we reiterate one of our central points. Problems in the natural environment are not isolated in nature. They are sustainability problems, meaning that they are part of a web of interconnected economic, social, and ecological dimensions. Understanding the web and the organization's role in it is a complex process that will require strategic managers to account for a host of multidimensional, rapidly changing success factors that are often novel and discontinuous, thus meeting all of Ansoff's four factors related to the changing nature of today's strategic problems (discussed at the beginning of this chapter). Thus, it is essential that organizations today develop new forecasting techniques, new analytical tools, and new thought processes that will allow strategic managers to fully understand and account for the economic, social, and ecological dimensions of the decisions they make.

Sustainability and Free Trade

As noted in the beginning of this chapter, the globalization of the economy has increased the turbulence experienced by organizations today. One of the most confounding macroenvironmental issues related to bringing sustainability into the strategic management formula involves the emergence of free trade. Citing the principles outlined in free trade agreements such as the General Agreement on Tariffs and Trade (GATT) and the North American Free Trade Agreement (NAFTA), many argue that free trade principles are incompatible with sustainability. A few years ago an anonymously penned poster mysteriously appeared in capitals all over the world portraying "GATTzilla," a horrendous monster devouring the planet by spreading wastes, destroying species, polluting water, and crushing democracy. On the other side of the argument are those who believe that free trade is, in fact, a primary requirement for a sustainable world economy. They argue that free trade provides developing countries with the opportunity to develop strong economies that will not have to rely on the exploitation of natural resources and cheap labor as the basic foundations of economic success. In this section, we will examine more closely the relationship between free trade and sustainability.

Free Trade and Global Competitive Advantages

Michael Porter proposes that the economic theories of comparative advantage that guided free trade for decades do not even address the right questions, much less provide the right answers.[9] These theories focus on understanding why some nations are more competitive than others, but Porter points out that nations do not compete. Instead, industries and industry segments are the competitors in the global market. The role of the nation is to serve as a supportive "home base" for the competitors. Thus, the proper question is, why do particular nations provide favorable conditions (what Porter calls a good home base) for the emergence of globally competitive firms and industries?

In answering this question, Porter dispels many fallacies related to the theories of free trade. For example, he points out that many nations are quite well off despite high trade deficits or high exchange rates for their currencies. He also shows that an abundance of cheap labor or natural resources is usually a poor mechanism for achieving advantages in the global marketplace. The advantages once provided by low wages and cheap natural resources are being replaced by the ability to adopt new technologies that bypass or reduce dependence on these factors. Porter maintains that nations that build their economies on the backs of cheap labor are not exactly ideal industrial role models. He believes that being successful despite high wages is a much more worthy ideal.

Porter identifies four determinants of a nation's ability to provide an effective home base for particular firms and industries. The first of these are factor conditions; a nation should be able to adequately create, upgrade, and specialize its human, physical, knowledge, and infrastructure factors to meet the needs of its industries. Of these, the most important are "advanced factors" such as sophisticated electronic communication capabilities, highly educated personnel, and a strong commitment to scientific research. The second determinant is the demand conditions that exist within the home base nation. Firms and industries that are required to constantly innovate and improve in order to be competitive at home typically will be better equipped to compete abroad. The third determinant is the presence of strong related and supporting industries (suppliers, software firms to support computer manufacturers, etc.), and the fourth determinant is how well the strategies and structures of the firms in the industry fit the dynamics of the global market as well as how much rivalry exists among the firms in the industry.

Porter makes two points about the sustainability dimensions of free trade in his writings. As mentioned above, he warns against basing competitive advantages on cheap natural resources and labor. He also argues that home base nations can provide global competitive advantages for firms by adopting strict environmental regulations that set ambitious targets and allow firms and industries to be innovative in the ways they pursue those targets.

Free Trade, Resource Efficiency, and Social Welfare

Another prominent scholar who has written extensively on the role of free trade in sustainability is Herman Daly.[10] Like Porter, Daly points out that the theories of free trade are woefully inadequate, and that this has led to several serious free trade issues. The first issue is the inefficient allocation of resources. Daly states that trade across the globe is very inefficient because it typically involves excessive transportation costs as products are produced on one side of the world and then transported to the other side for consumption. He humorously laments how much more efficient it would be if the Americans and the Dutch quit shipping cookies to each other and instead simply shared recipes. Second, Daly points out that free trade under the outdated principles of comparative advantage often leads to an unfair distribution of resources, pointing out that if there were truly nations with impermeable capital and labor markets, this would eventually lead to some nations' having absolute advantages over others as capital and labor followed lower production costs. Daly's third criticism of free trade is that it flies in the face of sustainability because there are no workable mechanisms built into the current free trade formula to account for the potentially broad differences in how environmental and social costs are internalized and externalized in different nations. Those nations that allow these costs to be borne by society and the ecosystem rather than including them in the price of goods and services will have significant cost advantages in a free-trade marketplace. According to Daly, these free-trade fallacies mean that wage rates for labor in high-wage countries will stagnate, that the cultural and economic bases of individual communities and nations will likely erode, and that ecological problems will be intensified.

Daly agrees with Porter that low wages and cheap natural resources are not an appropriate competitive advantage and that international trade takes place between individual firms and industries rather than between nations. He points to how massive debts accumulated by developing nations have forced them to overharvest rainforests and other natural resources to meet their financial obligations. However, Daly breaks with Porter by suggesting that ecological and social dimensions would be more prominent in international trade if trade were conducted between national communities rather than individual firms. He argues that nations could more effectively seek multilateral balances based on the physical amounts of resources traded (as opposed to financial balances). This would allow a nation to have resource deficits or surpluses with specific nations while maintaining an overall balance.

With regard to dealing with the social and ecological inefficiencies related to free trade, Daly argues that those nations that have regulations and other mechanisms designed to internalize social and environmental costs into the price of their products and services should be allowed to charge tariffs on imports from nations that do not. He points out that such tariffs would simply extend the already accepted free-trade practice of levying tariffs for dumping—selling products in other countries below actual production costs—to include full environmental and social costs as well as more traditional production costs. Whereas many would argue that tariffs such as those proposed by Daly constitute protectionism, he argues differently. He points out that protectionist tariffs are levied to protect inefficient industries from more efficient foreign manufacturers, but the tariffs he is suggesting are designed to assure that all costs, including social and environmental costs, are included in the definition of efficiency.

Free Trade and the Local Community

Michael Shuman believes that the patterns of corporate mobility that characterize free trade today pose an ominous threat to heretofore self-reliant communities across the globe.[11] Shuman discusses the perils for local communities when they become too reliant on mobile corporations for their economic base. He points out that because corporations today are so free to search the world for lower-priced resources, cheaper labor, and so forth, they are much more likely than they were in the past to close operations and move to areas where labor and materials are less expensive to acquire. Shuman identifies four ways that this threatens the stability and survival of communities. First, it causes a decline in the quantity and quality of jobs. Second, it imposes huge costs on all levels of government. Third, it contributes to the gradual decline of local culture, and fourth, it undermines the capacity of communities to plan for the future.

Many communities have seen their economies go into serious decline because employers have pulled out to find cheaper labor, cheaper materials, less government intervention, and so forth elsewhere. Take, again, the example of Johnson County, Tennessee (discussed previously in Chapter 2).[12] The county has sported a vibrant agricultural industry throughout most of its history, first with beans and more recently with tobacco. The county also attracted companies—such as Timberland Shoes, Sara Lee Clothing, and Levi-Strauss—looking for highly productive unskilled and semiskilled workers willing to accept relatively low wages for their work. Unfortunately, Johnson County cannot compete with the low wages being paid in many developing nations, so all three of these firms left Johnson County for cheaper labor outside the United States. Furthermore, tobacco sales have been declining for years and the U.S. Government Tobacco Allotment program is shrinking. Because of its economic problems, the county's hospital has closed, resulting in a chronic shortage of available health care, unemployment is high, about half of those who have jobs drive long distances to work, educational achievement levels are low, and most of the young people are fleeing the community if possible.

Shuman suggests that dealing with these community issues will require *going local*. Going local means not relying too heavily on mobile corporations, but rather finding ways to create and maintain locally owned businesses that use local resources, employ local workers at decent wages, and serve the needs of local consumers. If successful, such a strategy keeps more financial resources as well as human resources within the community. By keeping a greater portion of the economic wealth of the community circulating within the community, levels of employment can increase, local businesses can grow, income levels can rise, public welfare can improve, cultural heritage can be more easily preserved, and the natural environment can be better protected.

But what is really at the heart of successful going-local efforts? It is clear that, regardless of their strong desires and valiant efforts, the citizens of Johnson County, Tennessee, have a long row to hoe before they can expect to see much success from such efforts in their community. Their educational base and their infrastructure are weak, they have no significant knowledge labor pool, and they have historically followed a cheap labor strategy. Essentially, in Porter's terms, they have not created a sufficient home base to support their going-local efforts. Thus, before communities can embark on realistic going-local strategies, they need to build competitive home bases that provide solid foundations for those strategies.

Industry and Competitive Analysis

The industry (or competitive) environment is typically characterized by a group of firms producing the same or related products or services. Forces from the industry environment directly affect the firm, and the amount of influence the firm has over its industry is dependent on the dominance of its competitive position. Most strategic management books utilize Michael Porter's Five Forces Model as a framework for analyzing the competitive forces within the industry.[13] Like so many other models used to make strategic decisions today, the implicit assumption of this model is that the industry is operating within an economy closed to the greater society and ecosystem.

From the view of the Five Forces Model, industry analysis is traditionally portrayed in strategic management books from the rather static perspective of "what *is*" within the industry.[14] This model suggests that strategic managers scan the product market segments in which they compete for opportunities and threats without much regard for context. Their primary focus is on increasing market share within defined industry boundaries, and the competition is defined as those competitors who directly compete with them in individual product or service categories. Cooperative relationships are typically limited to those with direct suppliers and buyers. Capabilities to create value are viewed as residing in a single firm, and organizational performance is measured primarily in terms of how well the individual firm is managed with respect to its economic sustainability. Thus, within this traditional paradigm of industry analysis, strategic managers engage in adaptive learning within well-defined industry segments (see Figure 4.2).

Sustainable strategic management, however, requires a framework for industry analysis that reflects the firm's symbiotic relationship with the greater

Figure 4.2 **Traditional Industry Analysis of "What Is"**

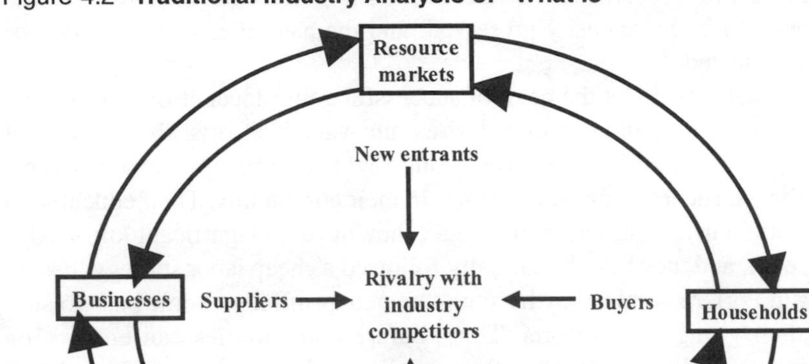

society and ecosystem. With a primary focus on competitive dynamics, traditional industry analysis virtually ignores the nonmarket context in which competition takes place. The biological concept of coevolution, however, (discussed briefly in Chapter 2) provides a framework to include society and the ecosystem in industry analysis.[15] The term "coevolution" was originally coined by Paul Ehrlich and Peter Raven to describe the reciprocal evolutionary relationships among entities.[16] In nature, organisms coexist in an ecosystem where each species has its own place or niche. Because their environment contains a limited amount of resources, the various species must compete for these resources. Species must also cooperate in order to survive within their specific ecosystem. Predator and prey interactions have coevolved into many strategies and counterstrategies in nature for both eating and avoiding being eaten. Through these interactions, species grow and change with each influencing the others' evolutionary development. These processes of reciprocal adaptation and development are the essence of coevolution.[17]

Whereas in traditional industry analysis, the structure of the environment influences the conduct of the firm, which, in turn, influences the performance of the firm, in coevolutionary industry analysis these three factors are seen as circular and interactive, meaning that the structure of the environment both influences and is influenced by the conduct and performance of the firm. In coevolutionary industry analysis, industry boundaries are blurred and the industries in which the firm competes are to a certain extent a matter of choice.[18] The industry is viewed as a community of coevolving firms that have coalesced around some form of innovation, such as the microchip. These firms are both competitors and alliance partners in a reinforcing system of symbiotic relationships. As in predator/prey interactions, these organizations develop strategies to compete (to eat) and to cooperate (to avoid being eaten). For example, General Motors has alliances with numerous competitors, including Toyota, Fiat, Suzuki, Isuzu, Saab, and Daewoo. In coevolutionary industry analysis, competition is not between firm and firm but rather between communities of firms sharing complementary products or services, similar processes, and similar approaches to the marketplace. Cooperation extends beyond direct suppliers and buyers to include all the participants in the community and in the relevant stakeholder and industrial ecosystem networks.

Sustainable strategic management's basic assumption that the firm has a symbiotic, coevolving relationship with the greater society and ecosystem requires that strategic managers analyze both the competitive dynamics of the communities of firms in which they participate and the other stakeholder networks and industrial ecosystem networks in which their firm participates. This provides a framework for sustainable strategic managers to analyze not only "what is," but also "what can be." Stakeholder relationships in the coevolutionary industry environment of today are more network-based than dyadic in nature. The stakeholders of firms today exhibit many direct and indirect interdependencies, resulting in the simultaneous influence of multiple stakeholder demands on the firm.[19] A recent survey of corporate executives by the Business Council for Social Responsibility noted a new attitude of openness and willingness to cooperate between firms and their stakeholders. The study's findings indicate that unlikely partnerships are developing between former adversaries as they recognize the need to work together as stakeholders with common interests.[20]

Industrial ecologies (discussed further in Chapter 8) are special types of stakeholder networks in which the firms in the network are committed to the core value of sustainability. Industrial ecologies are a set of interorganizational arrangements where two or more organizations attempt to recycle material and energy by products to one another as inputs. Whereas these industrial ecologies typically focus primarily on waste exchange arrangements for the purposes of pollution prevention and cost reduction, in sustainable strategic management, these networks would have an expanded focus that addresses resource depletion, overconsumption, biodiversity, community health, employment issues, and so forth.[21] Figure 4.3 illustrates coevolutionary industry analysis.

Figure 4.3 **Coevolutionary Industry Analysis of "What Can Be"**

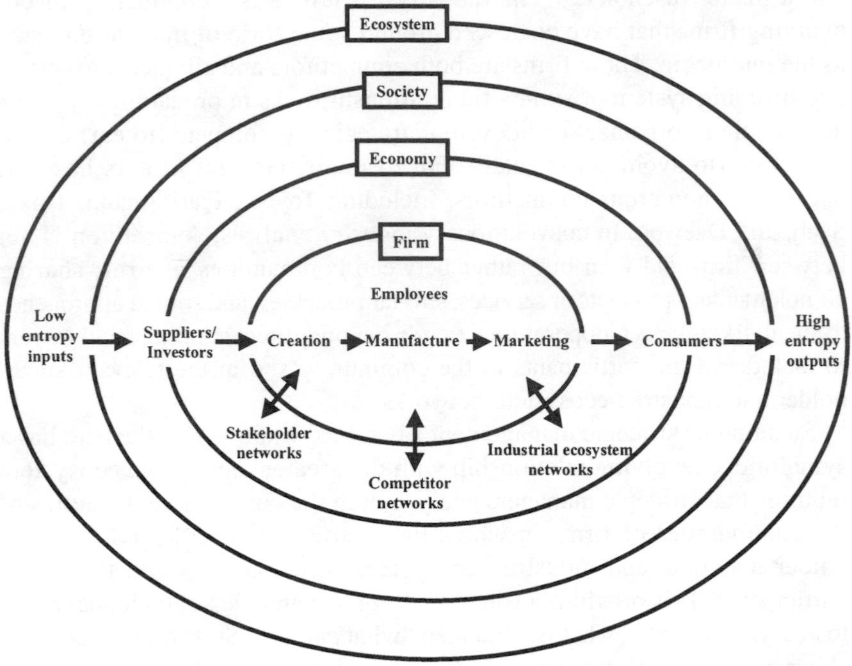

Ecosystem
Society
Economy
Firm
Employees
Low entropy inputs
Suppliers/ Investors
Creation
Manufacture
Marketing
Consumers
High entropy outputs
Stakeholder networks
Competitor networks
Industrial ecosystem networks

Environmental Forecasting

As previously discussed, the turbulence in the business environment is characterized by novel, discontinuous change, meaning that predicting the future based on the past is very difficult. This renders traditional forecasting techniques much less effective for predicting future trends unless they are combined with forecasting techniques that are not based on using the past as the sole predictor of the future. Strategic managers should be very cautious about limiting their forecasts to a single quantitative point estimate as the only possible future outcome in today's turbulent environment.

Forecasts shape strategists' view of their future and are integrated into the strategy formulation process. As mentioned in Chapter 1 and elsewhere, the ability to create the organization's future by thinking generatively, questioning established mental models in the formulation process, is a cornerstone of sustainable strategic management. Scenario building is a forecasting technique that creates a climate that fosters more generative, out-of-the-box learning within the strategic management team. Multiple scenario analysis (illustrated in Figure 4.4) has evolved over the past decades as a forecasting tool that allows strategic managers the flexibility to develop several paths to

Figure 4.4 **Multiple Scenario Analysis**

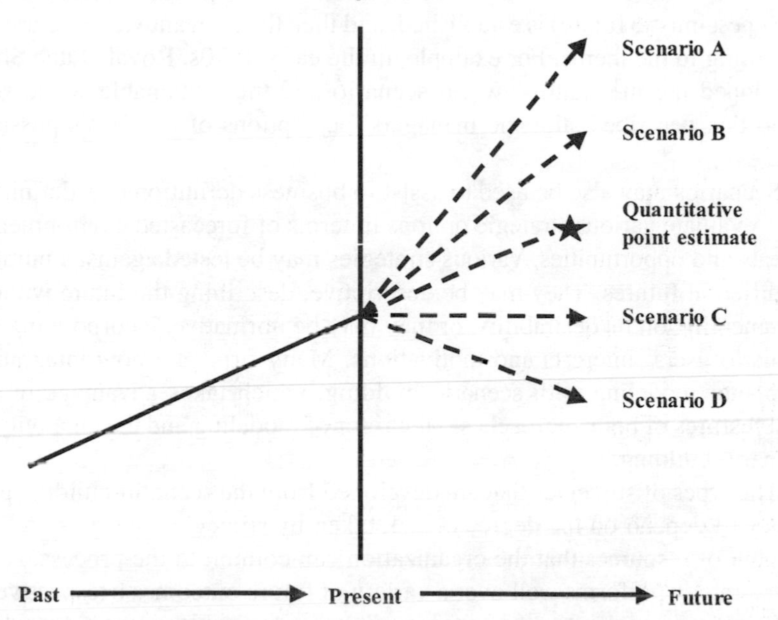

the future. The use of scenario analysis reflects the explicit recognition by strategic managers that the future is unpredictable. A scenario is a flowing narrative (rather than a quantitative point estimate) that depicts the general direction of change, where each scenario describes a possible path to the future. Thus, scenarios are stories based on mental models of the future. The use of scenario analysis was originally developed at Royal Dutch Shell during the late 1960s to prepare managers to think more clearly about the future.[22] Utilizing scenario analysis, Shell managers were prepared for the eventuality (not the timing) of the 1973 oil crisis and the global glut of oil in 1981 after the start of the Iran-Iraq War (1980–88).

Scenarios may be constructed for various purposes.[23] Scenarios may be anticipatory in that they help strategists anticipate and understand risk. Anticipatory scenarios (also referred to as inductive scenarios) start with identifying key environmental factors and their possible outcomes, and then scenarios are developed by utilizing a few of the plausible outcome combinations. Anticipatory scenarios search for possible causes that could lead to a given future state, placing more emphasis on goals. These scenarios seek more explanations than consequences. Another purpose of scenario analysis may be exploratory (also referred to as a deductive scenario), designed to help strategists discover strategic options of which they were previously unaware. Each scenario begins in the present, and the consequences are unfolded into the future. An overall theme (such as the most optimistic and most pessimistic future) is established, and then the relevant variables are set according to the theme. For example, in the early 1970s, Royal Dutch Shell developed the mercantilist world scenario and the sustainable world scenario that described strategic managers' perceptions of the firm's possible futures.

Scenarios may also be used to assist in business definition or redefinition or to evaluate various strategic options in terms of forecasted environmental threats and opportunities. Various strategies may be tested against a number of different futures. They may be descriptive, describing the future without commenting on its desirability, or they may be normative, incorporating the scenario users' interests and motivations. Many firms are now integrating computer modeling with scenario building, which takes advantage of the best features of both approaches—the rigor of modeling and the creativity of scenario building.

The types of strategies that are developed from the scenario-building process will depend on the degree of risk taken by strategic managers and the amount of resources that the organization can commit to the process. A robust strategy performs well over a variety of future outcomes irrespective of actual outcomes. This single strategy that is acceptable in most futures is conservative and risk adverse in nature. A flexible strategy is one that keeps options open until the future is clearer, and it entails a quick response rather than long-run anticipation. This strategy entails a great deal of risk. With a multiple-coverage strategy, a specific strategy is developed for each possible scenario. This strategy is quite comprehensive and requires a commitment of extensive resources for the scenario building process.

The most important outcome of the process of scenario building, however, is strategic manager learning and development.[24] The art of scenario building is an organic process where learning advances in an evolutionary fashion. The generation of scenarios provides a process that stimulates thinking about the future and creates a dialogue about strategic issues. Moreover, it builds a consensus based on the strategists' mental models of the future. The process brings to light inconsistencies in the mental models of strategic managers and provides an environment that enhances the ability to create a better future for the firm.

Conclusions

Given the uncertainty of an external environment characterized by turbulence, complexity, discontinuity, and change, strategic managers engaged in sustainable strategic management must develop a systems-based environmental scanning culture within their organizations. This means that macroenvironmental scanning must be expanded to include the social and ecological sectors and that traditional industry analysis must move toward coevolutionary industry analysis. The current models of industry analysis presented in most strategic management books assume away many of the current realities of the global marketplace and the home of all business activities, the planet earth. Within the context of sustainable strategic management, strategic managers will have to engage not only in adaptive learning as they respond to the demands of the marketplace but also in generative learning, questioning the fundamental assumptions and values on which their decisions are based. Strategic managers will have to work to rewrite industry rules and create opportunities beyond existing industry boundaries to include the larger society and ecosystem. This will allow them to focus on "what is" within their industries as well as to think in terms of "what can be" within and beyond their industries. Successful formulation of sustainable strategic management requires that strategic managers be able to create the future for their organizations by thinking and learning outside the box of traditional industry analysis, and scenario analysis provides an excellent tool for such generative processes.

5

Sustainable Strategic Advantage Analysis

Strategic management scholars agree that the goal of strategic advantage analysis is to determine the strengths and weaknesses of the firm, and, through analysis and evaluation, determine the core competencies on which the firm's competitive strategy will be based.[1] The resource-based view of the firm has become a popular framework to utilize in understanding how firms manage their internal resources to achieve a sustained competitive advantage. The process of resource assessment takes place at the functional levels of the firm and entails four stages. The first stage is a data gathering phase in which the organizational resources (both tangible and intangible assets and organizational capabilities) are profiled to determine what resources the organization currently has and what potential strengths and weaknesses those resources have for the firm. Some significant research has appeared over the past few years that has extended the resource-based view of the firm to the greater natural and social systems.[2]

The second stage involves analyzing and evaluating these potential strengths and weaknesses using both internal and external criteria. Internally, value-chain analysis, popularized by Michael Porter,[3] is utilized to disaggregate the firm's activities according to their value-creating functions so that the relevant resource or capability can be analyzed for its potential as a cost or differentiation strategic driver. A competitive advantage can be achieved only if the firm performs its value-creating functions at a lower cost than its competitors or in such a way that its products or services are so unique that consumers are willing to pay a premium price for them. Porter argues that each of a firm's activities—from physical creation through manufacture to the consumer—can be analyzed in order to better understand the firm's sources of costs and market differentiation. This analysis enables the firm to focus on adding value to its activities by finding ways to reduce costs and improve products at each stage of the value chain. The value chain includes primary activities, which are those directly related to manufacturing and marketing a product, and support activities, such as human resource management, information systems, auditing, and so on, that provide inputs so that the creation of products or services can take place.

Value-chain analysis encourages top managers to engage in strategic thinking rather than just performing an audit of the functional levels of the firm. The value chain can be conceptualized as virtual and capability focused rather than physical and product focused. In this view, value is created through a community of partnerships and alliances focused on innovative ideas that often require firms to span multiple industries. Working together, the community creates value and shapes its own future. The value chain, according to this perspective, is a system of value-creating processes that provide a unique value mix for the members of the community.[4]

Other internal criteria for the evaluation of potential strengths and weaknesses include a comparison to past performance and a comparison to the internal quality standards of the organization. From an external standpoint, the potential strengths and weaknesses are compared to the critical success factors in the product markets where the firm competes, the capabilities of the firm's strategic group members, and the industry dynamics.

The third stage of strategic advantage analysis entails the determination of the firm's core competencies—those strengths that are unique, nonimitable, competitively superior, and valuable. Core competencies are developed via functional level strategies and are the cornerstones of competitive strategy. Once core competencies are identified, the last stage of the process is to integrate them into the strategic management processes of the firm by exploiting the core competencies in the marketplace via the firm's competitive strategy. A sustained competitive advantage is the result of this process. Due to the environmental turbulence discussed in the previous chapter, a sustained competitive advantage has become more of a matter of the ability to change and adapt rather than one of location or position. The ability to think outside traditional industry boundaries, to manage the creative tension of cooperative and competitive relationships with multiple stakeholder networks, and to rewrite industry rules will be the necessary capabilities for organizational survival in the twenty-first century.

In addition to developing an understanding of the firm's resources, competencies, strengths, and weaknesses, successful strategic advantage analysis requires that firms understand their relationships with their stakeholders. Strategic managers can effectively assess the true nature of the competitive advantages and challenges they face if they can: determine who or what those persons, groups, and entities are that affect or are affected by the activities of the firm; assess the stakes that each of these has in the success or failure of the firm; and use this information to develop an understanding of stakeholder power, needs, and expectations. In other words, high levels of stakeholder management capability within the top management team will be a critical factor in successfully managing stakeholder networks in the current coevolutionary business environment.[5]

In the remainder of this chapter, we will explore strategic advantage analysis from the perspective of sustainable strategic management. We will examine the value chain in more detail and discuss how it can be adapted to better account for society and nature, and we will examine some of the key organizational stakeholders that are currently calling for improved ecological and social performance in business organizations.

Closing the Value Chain

As discussed above, Michael Porter's value chain has emerged as one of the most popular frameworks for analyzing and developing competitive advantages. From a sustainable strategic management perspective, the application of Porter's model is incomplete for at least three reasons. First of all, the stakeholder relationship implied in Porter's model is a narrow one, consisting of the supplier-firm-customer relationship, thus excluding the multiple stakeholder networks necessary for value creation.[6] It does not account for the value of resources in nature and the value of wastes after consumption, incorrectly assuming that the economy is a closed system. It also fails to account for the social capital embedded in the greater society, the human capital within the firm, and the multiple stakeholder networks that strategic managers must simultaneously manage. In sustainable strategic management, these factors are clearly important and should be included in value-chain analysis.

Second, the effective implementation of sustainable strategic management (covered in Chapters 8–10) will require value-chain activities that directly support sustainable strategic management, including human resource policies that enhance human capital, accounting systems that account for natural and social capital, design processes based on sustainability, and so forth.

Third, whereas strategic management texts typically depict the value chain as a straight line, sustainable strategic management will require that the chain be viewed as a closed cycle. The traditional straight-line depiction (see Figure 5.1) of the value chain certainly supports the old type I industrial ecosystems (discussed in Chapter 2) that convert virgin materials into products that are in turn discarded as wastes with little or no regard for the ecological or

Figure 5.1 **Type I Linear Value Chain**

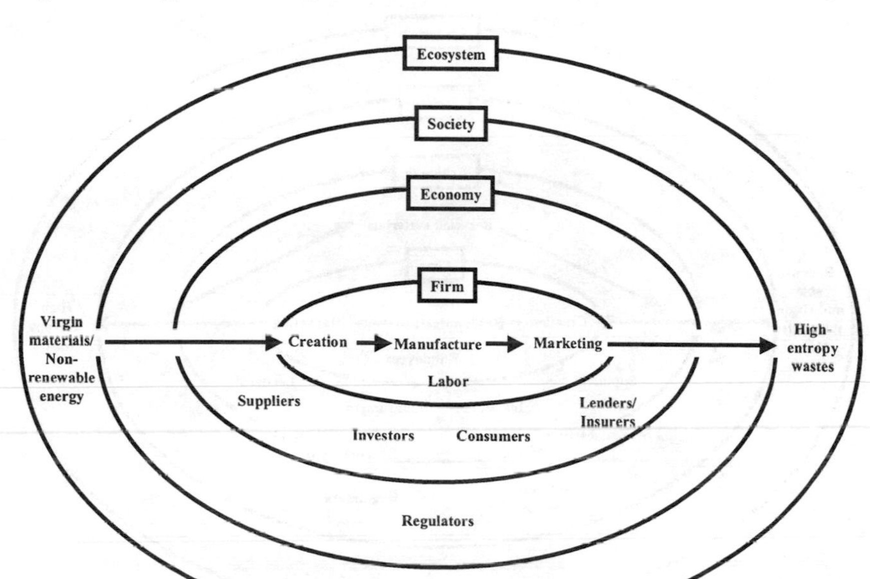

social impacts of this process, and it reinforces the narrow conceptualization that the firm's stakeholders consist primarily of suppliers and customers. In this straight-line depiction, the stake of employees is assumed to be accounted for merely by the fact that they work for the firm; they are simply labor, a factor of production.

By adding resources and wastes to the model but still depicting it as linear (see Figure 5.2), the model supports today's emerging type II industrial ecosystems that focus on making the old type I systems more ecologically efficient via resource reduction, process redesign, recycling, and reuse—often referred to as *ecoefficiency*.[7] Ecoefficiency creates value to the firm by enhancing cost and differentiating competitive advantages. Even though it accounts for high-entropy wastes and low-entropy resources, the linear depiction of the value chain operates under the old industrial paradigm of *cradle to grave*—resources from the cradle, wastes to the grave. These type II systems do less ecological harm than their type I counterparts, but they cause harm nonetheless.

In addition to opening the value chain to its ultimate resources and wastes, this model also opens the value chain up to the stakeholder influence of employees and the community, both of which have symbiotic, coevolutionary relationships with the firm. The human capital of the firm in this model is

Figure 5.2 **Type II Linear Value Chain**

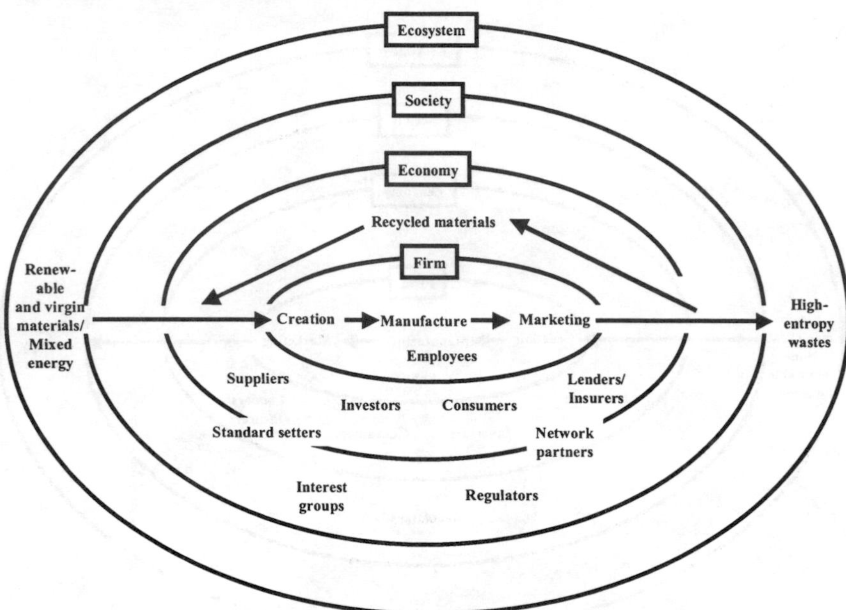

considered an important instrumental asset in the value-creating process. The development of human capital via innovative human resource practices that focus on the long-term competencies of the workforce is encouraged because it leads to higher levels of performance in organizations, increasing the value added at each stage of the value chain. Jobs in this model are designed to be economically, intellectually, and socially fulfilling, enhancing both the personal development of the human capital and the economic sustainability of the firm. This model also extends the value chain to community stakeholders—governments, activists, interest groups, and so forth—that can influence the firm's economic sustainability. Being a good corporate citizen in the communities in which the firm operates helps to build social capital and contributes to the long-run profitability of the firm. The ability to enhance both the social capital of the community and the human capital of the firm while contributing to the organization's economic sustainability is referred to as *socioefficiency*.[8]

Socioefficiency and ecoefficiency improve the state of the planet's natural and social capital as well as contribute to the firm's economic sustainability because strategic managers calculate the relative social and ecological costs and benefits before making decisions. Strategies based on eco- and socioefficiency are important incremental steps on the path toward sustainability. However, ultimately, achieving sustainability will require going beyond linear models. It will mean closing the value chain, transforming it from a straight line to a circle, and, in doing so, creating a mental model that encourages strategic managers to recognize the importance of granting stakeholder status to the greater society and ecosystem.

Closing the value chain into a cycle changes its orientation from cradle to grave to *cradle to cradle*, making it appropriate for the shift to a type III industrial ecosystem where renewable energy and resources are transformed into products whose wastes in turn serve as inputs for other biological or industrial cycles—referred to as *ecoeffectiveness*. Can the loop really be closed? Are type III industrial ecosystems really possible? William McDonough and Michael Braungart think so.[9] According to them, the key to making the paradigm shift from cradle to grave to cradle to cradle lies in redesigning industrial systems from those that defy natural metabolic processes to those that mimic natural metabolic processes. They point out that nature is not so much efficient as effective. Many of nature's processes have built in inefficiencies, but whatever wastes are generated are always absorbed and reused in the system. No wastes are left as wastes. They maintain that this should be the model for designing new industrial systems. To accomplish this, they suggest that products be designed to include only two types of materials: *biological nutrients*, materials that biodegrade and can be returned to the biological cycle, and *technical nutrients*, materials that do not biodegrade but can be circulated continuously through the industrial cycle.

Closing the value chain also requires that the absolute rights and limits of human capital and social capital be accounted for in value-chain analysis (see Figure 5.3). Within the closed-loop value-chain model, employees, the human capital of the firm, are no longer viewed as merely instrumental in value creation, but are considered to have intrinsic value in and of themselves. They become ends rather than means to ends.[10] Furthermore, the concept of the firm's social capital, like natural capital, is characterized by the irreversibility of its deterioration, the nonlinearity of its processes and its nonsubstitutability. For example, the loss of an indigenous culture, just as the loss of a species, is irreversible. *Socioeffectiveness* is the term used for such stakeholder awareness.[11] As with ecoeffectiveness, socioeffectiveness is based on the assumption that the economy is an open subsystem of the greater society and ecosystem, expanding the community of stakeholders to include nature. This encourages a stakeholder perspective that requires strategic managers to expand their paradigms to include the global community and the long-run importance of social and human capital in the value-creation process. Issues such as child labor, human rights, economic justice, the exploding human population, disease and poverty in the southern hemisphere, the overconsumption of resources and energy in the northern hemisphere,

Figure 5.3 **Type III Closed-Loop Value Chain**

Ecosystem
Society
Economy
Technical nutrients
Firm
Biological nutrients/ Renewable energy
Creation
Manufacture
Marketing
Biological nutrients/ Heat
Employees
Suppliers
Lenders/ Insurers
Investors
Consumers
Standard setters
Network partners
Interest groups
Regulators

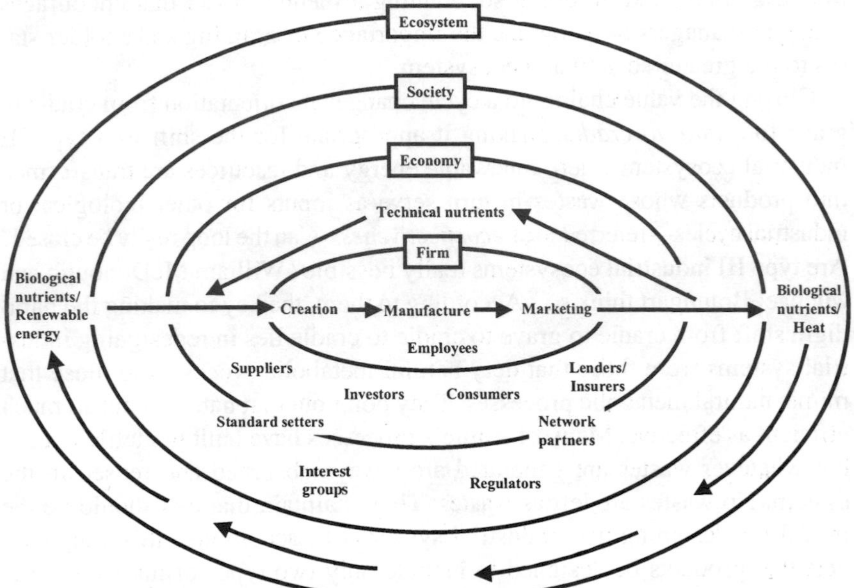

preservation of indigenous cultures, global warming, and so forth will likely be on the strategic agendas of these managers.

Before we end this section, we need to ponder one last question: Can firms that take a closed-loop value-chain perspective actually expect to achieve any meaningful competitive advantages or to improve their strategic capabilities for their efforts? Can they really expect that their efforts will create value for their firms? Given that no pure type III industrial ecosystem currently exists, answering this question definitively is not possible. However, there is a preponderance of evidence strongly suggesting that within the type II value chain, cost savings and market differentiation advantages are certainly achievable via processes that help to account more completely for natural resources and wastes, and one empirical study in particular found that firms following proactive environmental management strategies could significantly improve competitive capabilities for stakeholder integration, higher order learning, and continuous innovation.[12] These capabilities are necessary to provide the skills for strategic managers to question the fundamental assumptions of how they are doing business and to integrate the complex stakeholder demands of including the greater social system and natural environment into their strategic management processes. The type III closed-loop value chain provides the conceptual framework for strategic managers to examine their strategic alternatives and to ask themselves, "What can be?"

Nature's Stake in Organizations

So, in the final analysis, how much weight do strategic managers give the greater social system and ecosystem as they assess their strengths and weaknesses and determine their competitive advantages? Just how much consideration does society deserve, does nature deserve? As we discussed above, we believe that answering this question begins with the recognition that the earth is a stakeholder of significant import for business organizations. Recognizing this is critical for integrating sustainability into the strategic core of organizations because it stresses the key point that the long-term survival of business and the long-term survival of the earth are intricately interconnected.

The belief that the earth is a stakeholder is shared by many scholars.[13] However, some have criticized this notion. One of the most compelling criticisms is that including the earth as a stakeholder changes the definition of the term too drastically. In a narrow sense, the definitional criticism has some merit. Freeman's classic definition of stakeholders, "Persons or groups that can affect or are affected by the achievement of the organization's objectives," certainly says nothing about planets or other nonhuman entities.[14] From a broader perspective, however, the earth includes and supports all of humankind within its sphere, giving it a significant "person or group" dimension. Moreover, the earth clearly meets the rest of the definition of a stakeholder as set out by Freeman; it "can affect or be affected by the achievement of the organization's objectives." It is the home of humankind, the geographical location of all business activity, the source of the resources and energy necessary to make the economic engine purr, and the sink into which the wastes of economic activity are poured. From this perspective, the earth is not just an important stakeholder in the strategic decisions of business organizations. It is the ultimate stakeholder.

Another criticism of assigning stakeholder status to the earth is that the planet itself cannot attend shareholder meetings, sit on boards of directors, or otherwise participate in the activities of organizations. Even though the earth certainly cannot physically sit in the chairs or attend the meetings, it has some powerful human friends who certainly can and do. And these friends can be found not only in boardrooms and meeting rooms but also in retail stores, financial markets, courtrooms, the media, the halls of government, and in organizations themselves. In fact, nature has a large cadre of stakeholders representing its interests in the immediate business environment, including consumers, investors, suppliers, employees, legislators, regulators, interest groups, lenders, insurers, and environmental standards setters. The prospects for tying both short-term and long-term organizational success to satisfying the needs of these stakeholders give the planet plenty of clout in the organization. This is likely why firms wishing to take a proactive approach with regard to their environmental performance consider these stakeholders so important and worth heeding.[15] Below we examine in more detail some of the stakeholders that are effectively representing the earth in business organizations.

Regulators

There was some early U.S. government attention focused on social and environmental ills. President Abraham Lincoln put an end to slavery with his Emancipation Proclamation, Theodore Roosevelt made conservation a popular political issue, workers were given rights to organize, collectively bargain, and strike, and business organizations were held liable by the courts for employee health problems resulting from exposure to asbestos and coal dust. However, prior to the 1960s, most government regulation aimed at business focused on economic activities—antitrust legislation, pricing legislation, and so forth—with little government attention paid to social and environmental issues. The primary reason for this lack of attention to social and environmental agendas was that society and business were on a honeymoon of sorts. During the postwar 1950s, there was pent-up societal demand for the consumer goods and services that had been restricted due to the war effort. Thus, during this period the goals of society and business were congruent: the accumulation of wealth. Strategic managers' jobs were relatively easy ones; all they had to do to be successful was to position their products and services in the correct market segments and execute the strategy. The primary critical success factors were related to finding competitive product-market-technology fits, and the primary variables considered in strategy formulation were market variables.[16]

However, in the 1960s, the goals of society and business began to diverge because of changing social values and expectations of business. The rise of social and environmental activism in the United States during that time raised awareness of social and environmental problems and planted values such as social justice and conservation firmly into the public consciousness. Once these values were implanted in society, they moved rapidly through the public issue life cycle and were translated into legislation, government policy, and so forth. Thus, from these changes in values a new form of regulation emerged addressing societal and environmental objectives. Since that time, there has been an explosion of social and environmental regulations, policies, agencies, court decisions, and so forth that target business organizations' relationships with the greater society and ecosystem to which they belong.

The 1964 Civil Rights Act, reflecting the emerging social value that economic opportunity is a birthright for all citizens regardless of heritage, is the most sweeping social legislation of all time in the United States. The act created the Equal Employment Opportunity Commission, which monitors business behavior to ensure that all citizens are afforded their rights to economic opportunities regardless of race, creed, or color. In 1965, President Lyndon Johnson signed Executive Order 11246, which required all government contractors to take affirmative actions to accelerate the movement of minorities into the workplace. Recently, affirmative action has come to refer to programs that give members of certain groups preference in determining access to positions from which they were formerly excluded. Through this legislation and subsequent case law development, society was communicating to the business community its demand for products and services produced in a setting of equal opportunity.

During the 1960s, strong social values arose regarding the safety and health effects of the products and services produced by business organizations. Ralph Nader's 1965 book, *Unsafe at Any Speed* (New York: Grossman), exposed the dangers of Chevrolet's rear-engine Corvair and served as an early impetus for the consumerism movement in the United States. This movement led to the passage of the Consumer Product Safety Act, which created the Consumer Product Safety Commission (CPSC) to protect consumers from the risk of injury and deaths from the use of consumer products. The CPSC and the Federal Drug Administration are the two primary regulatory bodies protecting public health and safety. Safety also arose as a workplace issue in the 1960s, resulting in the 1970 legislation creating the Occupational Safety and Health Administration (OSHA) to monitor organizations with respect to worker safety and health.

With the globalization of the firm's markets (discussed in the previous chapter), there has been increased concern for human rights at the international level. The concept of human rights implies that each person has a right to respect and dignity on the basis of their humanity. This concern led to two new covenants being added to the Universal Declaration of Human Rights, adopted by the General Assembly of the United Nations in 1948. The Covenant on Civil and Political Rights and the Covenant on Economic, Social, and Cultural Rights were adopted in 1966. These covenants are designed to more clearly specify all human beings' rights to freely choose their religion, to freely associate with others, to be free from torture, slavery, and discrimination, to pursue an education, to have just and favorable working conditions, and to participate in cultural life. Another area of rights has recently emerged onto the world stage concerning issues requiring international cooperation, such as environmental protection and the right to development.[17]

Rachael Carson's *Silent Spring* (Boston: Houghton Mifflin, 1962) described a time when the birds are silenced as a result of the unregulated use of herbicides and pesticides. Her book had a tremendous impact on society's awareness of and values about protecting the natural environment, and was followed closely by the environmental movement in the United States. Eventually the Environmental Protection Agency (EPA) was created in 1970, with the mission of coordinating various environmental efforts and developing comprehensive environmental policy at the national level. In addition to the EPA, there is currently a conglomeration of federal agencies in the United States responsible for generating and enforcing environmental regulation. The Department of Energy (DOE), the Department of Justice (DOJ), the Occupational Safety and Health Administration, the Federal Trade Commission (FTC), the U.S. Forest Service, the Bureau of Land Management, and the Fish and Wildlife Service are some of the other U.S. agencies that are responsible for environmental regulation.

There are over sixty U.S. federal laws that mandate environmental regulations. Many focus on specific environmental problems, such as air pollution, water pollution, asbestos, species protection, and so on. Others focus on specific industries, such as agriculture and utilities. Others focus on regions or habitats, such as coastal zones, forests, and wetlands. Still others focus on specific organizational practices, such as the accuracy of environmental claims made in advertising. Of these sixty or so laws, six are generally considered to constitute the primary core of U.S. environmental regulation. These are: the Clean Air Act (CAA), the Comprehensive Environmental Response, Compensation, and Liability Act as amended by the Superfund Amendment and Reauthorization Act (CERCLA/SARA—better known as the Superfund law), the Clean Water Act (CWA), the Resource Conservation and Recovery Act (RCRA), the Safe Drinking Water Act (SDWA), and the Toxic Substances Control Act (TSCA).

Individual states also have significant influences on the environmental behaviors of business organizations. For example, California has the toughest air pollution standards in the nation. The California Air Resources Board has instituted automobile emission standards that get progressively tighter over a twelve-year period, and many eastern U.S. states have adopted some version of this California law. Other states have focused on other issues; for example, a number of states now have strict packaging laws that regulate types of packaging, deposits on packaging, etc. Furthermore, regulators in the United States generally have teeth. Firms can find themselves paying large penalties, spending millions on cleanups and restoration, and, in some cases, having their executives facing potential jail time. For example, two important cases over the past fifteen years involving RCRA violations—the

United States v. Dean and the *United States v. Hoflin*—ruled that executives could be held criminally liable for the environmental crimes of their subordinates.[18]

Environmental regulation certainly goes beyond the U.S. border. Germany has adopted some of the toughest and most creative environmental laws and policies in the world. Of special interest is the German Packaging Legislation. Under the law, organizations are required to accept used packaging from consumers and recycle it unless technically infeasible to do so. In July 1994, Germany passed another landmark environmental law, the Closed Substance Cycle and Waste Management Act, which essentially extends the idea of the closed-loop management of packaging established by the packaging law to include all types of wastes generated in the German economy. Wastes, according to this law, are broadly defined as anything unrelated to the original intent of a product that arises as a result of production, processing, or consumption. This definition encompasses everything from toxic by products of production to used cars and old newspapers. The law specifies that it is no longer economically or ecologically viable to maintain a dual system in which industry produces wastes and then leaves it up to the public to find ways to dispose of it. Rather, the law is based on the polluter-pays principle, which says that those responsible for generating wastes are responsible for finding ways to avoid generating the waste in the first place and to recycle, reuse, or safely dispose of them.

Not only individual nations are getting involved in regulating environmental performance, so are the major international trade associations, such as the European Union (EU) and the North American Free Trade Agreement (NAFTA). The Single European Act of 1987 mandated that all EU citizens have the right to live in a clean environment, so a central theme of EU policy over the years has been the tying of environmental performance to the union's economic strategy. One example of EU environmental regulation is the European Eco-Management and Audit Scheme (EMAS). EMAS became EU regulation in 1992, providing a framework for firms to assess their environmental impacts and plan future actions. NAFTA, after much debate and renegotiation, was finally passed with its environmental side agreement. This puts into place an environmental philosophy in NAFTA similar to that in the European Union. Officially known as the North American Agreement on Environmental Cooperation (1993), the NAFTA environmental side agreement affirms that conservation, protection, and enhancement of the environment in Canada, Mexico, and the United States are important, and that the three nations will cooperate in sustainable development efforts designed to bring well-being to present and future generations.

In addition to the trade association agreements, there have been many international treaties negotiated to regulate environmental problems across borders. In 1992, the United Nations Conference on the Environment and Development, the so-called Earth Summit, held in Rio de Janeiro, resulted in nations of the world signing agreements committing themselves to developing regulations and incentive programs that address a wide variety of the planet's social and environmental ills. The 1992 conference was followed ten years later by the World Summit on Sustainable Development in Johannesburg, South Africa. During the Johannesburg conference, the tenets of the 1992 conference were largely reaffirmed. The delegates committed the United Nations to seeking economic, social, and ecological balance, to eradicating poverty, to changing unsustainable patterns of consumption and production, and to incorporating sustainable development more effectively into the United Nations' overarching governance framework.

Two other important international agreements are the Montreal Protocol and the Kyoto Protocol. The Montreal Protocol, signed in 1987 and revised in 1990 by dozens of industrialized nations (including the United States), set standards for protecting the ozone layer, including phasing out the production of chlorofluorocarbons (CFCs). The Kyoto Protocol, drafted in 1997, is designed to reduce emissions of greenhouse gases worldwide in order to eliminate or reduce the prospects for disastrous climate change. Unfortunately, getting the Kyoto Protocol ratified has been somewhat of a struggle thus far because it has yet to garner a critical mass of signatories among nations that generate high levels of greenhouse gases. The nation that generates the most greenhouse gases and the one that has been most resistant to signing the agreement is the United States, and this resistance has slowed the ratification process significantly.

The most serious problem related to environmental regulations may be their complexity and resulting high costs. The fact is that organizations today are faced with an alphabet soup of often conflicting command-and-control regulations that cost them time, money, and, in many cases, restrict them from taking more effective actions to improve their environmental performance. Discussing this issue at length, Porter and van der Linde compare and contrast the regulatory environments of the United States and Scandinavian pulp and paper industries.[19] They point out that the lack of sufficient phase-in periods and technological flexibility in U.S. regulations means that pulp-and-paper firms are required to rapidly adopt expensive end-of-pipe technologies without the opportunity to develop any innovative process improvements that would allow for more economically feasible pollution prevention approaches. In Scandinavia, on the other hand, the regulations do not restrict technologies and include phase-in periods; this has led to the development of innovative pollution-prevention technologies in pulping and bleaching. These new technologies provide a significant competitive advantage for firms in the Scandinavian pulp and paper industry. Porter and van der Linde conclude that the important issue is not regulatory strictness, but regulatory structure. They maintain that strict regulations are necessary because they protect the natural environment and provide competitive advantages for organizations. However, they believe that regulations should be structured so as to promote product and process innovations that give firms first-mover advantages in the global marketplace. They define good environmental regulations as being strict, focusing on outcomes rather than technologies, allowing phase-in periods, and using market incentives rather than penalties. They believe that regulatory processes need to be more predictable and less costly, involving more industry participation and more technically capable regulators that can be helpful to industry.

Many are echoing Porter and van der Linde. The consensus is that there is a need for strict environmental regulations, but that regulations need to be more consistent, less intimidating, and more economically feasible for business organizations. They need to be less punishment-based and more incentive-based, and they need to give organizations some discretion in using their own creative processes to find ways to effectively improve environmental performance. Regulators also need to be more supportive, giving financial and technical support when necessary, and taking on the role of educators, communicators, and consultants. Regulators in the United States seem to be taking conclusions like these to heart. Recent U.S. regulatory efforts have been more incentive based, setting targets instead of requiring specific technologies, and so forth. For example, the EPA's Project XL, which began in 1995, allows firms with good environmental records to replace existing regulatory requirements with their own creative approaches to improved environmental performance.[20]

Consumers

Over the past several years, consumers have become more informed about the health, safety, social, and environmental impacts of the products they buy. The result has been a rise in the number of consumers who demand environmentally sensitive, socially responsible, safe products that can be reused or recycled rather than discarded as wastes. They want products that are high quality and durable, made with nontoxic materials, produced and delivered using energy efficient processes, packaged in small amounts of recyclable material, not tested on animals, not derived from threatened species, and not produced by child or forced labor. Products of this type not only are safe for the earth and its inhabitants but also serve as personal statements about the values of the consumers who purchase them.

Many surveys over the past decade or so have indicated that a majority of consumers are at least somewhat concerned about the impact of products on society and the environment. However, the actual buying patterns of consumers tell a slightly different story. For example, consumers often say they want socially and environmentally responsible products but then resist paying the premium prices often attached to such products, and they often refuse to inconvenience themselves in any way to acquire these products (i.e., they will not make a trip to a special store). Many think that the primary reason for the discrepancies between the words and deeds of socially and environmentally responsible consumers centers on the varying levels of commitment to social and environmental responsibility found among the consumer population. Generally consumers in the United States fall into three categories with regard to their level of commitment to buying socially and environmentally responsible products. There are those who are hard and fast in their commitment and willing to put their money and their actions where their sentiments lie. There are those who fit into a swing group; while these consumers talk a good game, they often let economic and convenience issues outweigh their social and environmental concerns in the marketplace. Finally, there are those who are either unconcerned about or hostile to social and environmental concerns.[21]

Another reason consumer buying patterns may demonstrate less commitment to environmental and social responsibility than expected is that consumers often do not trust the information they get from organizations about the social and environmental benefits of their products and services. Many consumers believe that these claims are little more than marketing gimmicks. Thus, trying to be socially and environmentally responsible becomes filled with choices that are complicated and that must be made based on data that are confusing, incomplete, and often misleading. Much of the problem stems from some highly visible cases of fraudulent social and environmental claims (e.g., biodegradable plastic garbage bags) that make people skeptical about all such claims. These misleading claims lead consumers away from organizations to other sources of information on the environmental impacts of products and services. Often they seek information from sources in the environmental community. These are the very sources that firms criticize for antibusiness, proenvironmental bias in the information they distribute. To alleviate this problem, firms have to be willing to provide accurate, honest information about the environmental and social performance of their operations and products.[22]

Consumers are also concerned about the safety of the products they buy. Through a series of legal developments and changing societal expectations, consumers have demanded that businesses be held more accountable for product safety. Today consumers are concerned about a variety of issues ranging from food scares such as mad cow disease to the flawed design of children's toys. This increased societal concern about product safety has led to the emergence of product liability as a monumental consumer issue. In fact, the concept of product liability has been expanded by legal rulings to include the doctrine of strict liability that holds every component of the value chain liable for harm caused to the user if the product sold is unsafe or defective.[23]

While it is good news for the planet and its people that consumers are becoming more socially and environmentally aware, a few words of caution are necessary here. First, identifying environmentally and socially responsible products is no easy task. Second, not all of the so-called environmentally and socially responsible products on the market live up to their claims; consumers need some way to separate truly responsible products from mere marketing schemes. Third, the term "environmentally and socially responsible consumer" is, in a sense, an oxymoron. Environmental and social responsibility goes well beyond simply buying more earth-friendly and people-friendly products. That would mean buying more in order to use less, a conundrum for sure. In fact, one of the principles at the heart of sustainability is to consume less, period.

Investors

Socially responsible investing, or *ethical investing* as it is often called, refers to the practice of screening investment alternatives based on social, environmental, political, community, or moral performance criteria. The idea of using investment power for social as well as financial returns dates back to the 1920s, when various religious groups demanded that their money not be invested in *sin stocks* (liquor, tobacco, and gambling). In 1928 the Pioneer Fund of Boston began eliminating from its portfolios companies that had operations in any of these three areas. Since then, many funds have emerged that screen their portfolios along numerous ethical, social, and environmental lines. Religious organizations have continued to be major influences in the ethical investing movement; for example, in 1962, the Christian Scientists began the Foursquare Fund, and in 1968, the United Methodist Church withdrew its $10 million portfolio from First National City Bank (now CitiGroup) because of the bank's investments in South Africa.

The social turmoil experienced in the United States in the 1960s was a catalyst for rapid growth in ethical investing. Social activist Saul Alinsky garnered support from hundreds of Eastman Kodak shareholders in 1966 (the group controlled almost 40,000 shares of voting stock), demanding that the firm hire 600 additional minority employees. Civil rights leaders lobbied

Chase Manhattan and First National City Bank to stop them from investing in South Africa. Stockholders of both Honeywell and Dow Chemical used annual meetings as public forums to air their objections to the antipersonnel weapons, napalm, and defoliants (Agent Orange) these firms were producing for use in the Vietnam War. In 1968, a Jewish synagogue requested that Alice Tepper Marlin, a securities analyst in Boston, develop a *peace portfolio* for it. When she had developed the portfolio, Marlin convinced her firm to run a newspaper advertisement offering the fund to other investors; some 600 investors responded to the ad. This interest in the peace portfolio eventually led to the formation of the Council on Economic Priorities (CEP) in 1969, which provides social performance ratings for individual companies. CEP evaluates corporations on criteria such as environmental performance, charitable contributions, the advancement of women and minorities, animal testing, community outreach, and family benefits.[24]

As the above discussion suggests, there are two basic approaches to ethical investing. The first involves investing in firms or mutual funds that demonstrate a sincere effort to be socially and environmentally responsible. Implicit in this approach is that firms with poor social or environmental records be avoided or withdrawn from investment portfolios. There has been a tremendous increase in the amount of money invested in this way. Twenty years ago, only about $40 billion were invested in ethical investments in the United States, but that figure has climbed to over $2 trillion today, about 13 percent of total current U.S. investments. However, are ethical investments as good for the investors' wallets as they are for their consciences? Data suggest they are. Trends over the past ten years have shown very reasonable returns on ethically invested stocks when compared to conventional investments. For example, during the first seven years of the 1990s, the Domini 400 Social Index (an index of socially responsible stocks in the United States) grew 302 percent while the Standard and Poor's 500 rose 262 percent.

The second approach to ethical investing involves the use of proxy rights and shareholder resolutions to initiate changes in the social and environmental practices of corporations. The aforementioned shareholder actions at Eastman Kodak, Honeywell, and Dow Chemical began a flood of such activities on behalf of the environment and other social causes. Two shareholder resolutions presented at the 1970 General Motors annual meeting resulted in the company's appointment of the first Black to General Motors' board, Lewis Sullivan (author of the well-known Sullivan Principles, guidelines for companies doing business in South Africa). From that point, the number of ethical proxy proposals at General Motors shareholder meetings increased exponentially; there were 111 such proposals presented at the 1983 meeting. Today about 120 institutions and mutual funds have leveraged close to $1 billion in investment assets in order to support shareholder resolutions related to the social and environmental performance of organizations.[25]

Ironically, probably the greatest impetus for the growth and maturity of proxy proposals related to improving environmental performance is also one of the most serious environmental disasters of all time—the *Exxon Valdez* oil spill. Exxon's lax attitude toward the spill and its cleanup resulted in the formation of the Coalition for Environmentally Responsible Economies (CERES), a consortium of investment institutions, environmental advocacy organizations, labor unions, public interest groups, and community activists. The purpose of CERES is to encourage corporate environmental responsibility. CERES's primary vehicle for accomplishing its purpose is the CERES Principles (originally the Valdez Principles). This is a set of broad principles that deals with biosphere protection, resource sustainability, risk reduction, product safety, damage compensation, disclosure of environmental mishaps, appointment of environmentalists to boards of directors and management positions, and annual self-audits of environmental activities. CERES uses proxy proposals to encourage business organizations of all sizes and genres to sign these principles, thus endorsing their long-term commitments to the process of achieving environmental sustainability. Today there are over seventy CERES signatories, including Ford, General Motors, Nike, Coca-Cola, Polaroid, and Timberland.[26]

Investors are currently experiencing a crisis in confidence in business organizations because of a handful of CEOs who have been charged with using improper accounting procedures to hide losses and overstate profits and earnings, bilking their companies, employees, and investors out of billions of dollars. Not only have investors lost confidence in the companies in which they invest, they have also lost confidence in the advice they receive from analysts on Wall Street. Society's outrage over the recent corporate fraud and lack of ethics of corporate management is being translated into a demand for prosecutions and new regulatory reform aimed at controlling ethics in the investment community. This adds a whole new dimension to the idea of ethical investing.

Employees

Employees are central to any discussion about organizational stakeholders. Employees are often the first to either suffer or benefit from the economic, social, and environmental performance of the firm. Employees are the central building blocks of the firm (its human capital), and their value to the firm is multidimensional. Employees are a diverse group of people whose interests and activities generally extend beyond their jobs; they are members of activist groups, they serve on community boards, they are involved in political activities, and they are members of professional, trade, and service organizations. Employees are often owners and stockholders of the firms that employ them. As such, employees often have an influence on or are influenced by the actions of the organization at levels well beyond their performance at work.[27]

One of the first strategic managers to declare his employees stakeholders was Ernest Bader, founder in the 1930s of the Scott Bader Company in the United Kingdom. In 1945, he declared, "The stakes of owners and customers in a business are temporary, transient and partial, but the employee normally . . . seeks in and through [the organization] a much wider personal satisfaction."[28] Bader proved his overriding commitment to his employees in 1951, when he gave them the company, making the Scott Bader Commonwealth the first employee-owned firm in the United Kingdom. Its operation is based on a code that stresses the organization as a working community, measures success of the organization in terms of technical, social, and political as well as economic dimensions, makes it the express duty of managers to make all jobs personally fulfilling for employees, mandates that decisions in the firm be made by consensus, and states that all employees are responsible and accountable for making decisions about their jobs. The code also includes commitments to produce only beneficial products, to protect the natural environment, and to question any activities that appear to waste the earth's resources.

Until his death in 1977, E.F. Schumacher, a friend and confidant of Ernest Bader, helped Bader to transfer the Scott Bader Company to its employees and served on the company's board. Like Bader, Schumacher believed that employees were the most critical stakeholders. Schumacher was very passionate about the meaning of work. He believed that the purpose of work was to allow people to develop their highest faculties, to work with others to accomplish common goals, and to provide goods and services that improve the quality of life. He said, "[People] should be taught that work is the joy of life and is needed for development, but that meaningless work is an abomination." His philosophy strongly supports the type III value chain assumption that the human capital of the firm has intrinsic worth. Schumacher believed that the lack of available meaningful work in our bureaucratized, mass-production society was a primary contributor to economic and ecological ills. He felt that when there is no intrinsic fulfillment from the job, the natural tendency is for employees to focus on getting more money and more things, which contributes both to inflation and environmental degradation.[29]

Employees are now operating under a new social contract in a high-risk environment that has required them to relinquish the job security that they had taken for granted in past years.[30] The new social contract requires employees to take more responsibility for their success in the employment relationship. Their compensation, advancement, and job security now depends on how much value they create for the firm rather than on their tenure or loyalty. Within the context of this new social contract, the employee rights movement has gained momentum. Employee rights are seen as employees' legitimate and enforceable claims to some desired treatment, situation, or resource. The most important employee rights are the right to be dismissed only with just cause, the right to due process and fair treatment, the right to freedom of speech and expression, the right to privacy, and the right to safety and health. Employee issues such as drug testing, the collection and use of employee information by employers, employee monitoring, whistle blowing, and so forth are currently of major concern to many employees. For example, after September 11, 2001, many organizations started performing background checks on all of their employees, which resulted in numerous layoffs. Employee rights advocates have questioned whether this was a violation of these employees' rights.

Recognizing employees as critical stakeholders is especially important for effective formulation and implementation of sustainable strategic management. As we discussed briefly in Chapter 1 (and will discuss further in Chapter 9), sustainable strategic management requires organizational structures and processes for managing transformational change. Such structures and processes require that employees constantly be encouraged to think out of the box, be adaptable, stay involved, and strive for innovative solutions. For example, total quality environmental management (TQEM), a standard process for achieving ecoefficiency in organizations, cannot be successfully implemented unless the human capital of the firm is directly and creatively involved in the process at all levels.[31]

Suppliers

Organizations interested in integrating sustainability into their operations, products, and services cannot do so in isolation. Socially and environmentally sensitive products and services can be produced only with socially and environmentally sensitive inputs. No product can be toxin free, organic, and safe for the environment unless the ingredients or parts that go into the product meet these criteria. Firms can treat their employees with all of the respect, dignity, and fairness possible, but if they are supplied with inputs made by workers who are poorly paid and poorly treated, they can make no claims about social sensitivity in their products. Thus, sustainability-based supplier relationships are critical for firms seeking to practice sustainable strategic management.

There are several prominent examples of firms that build their social and environmental reputations on the relationships with their suppliers. Starbucks has made it a practice to buy coffee beans from indigenous peoples and to pay them much more than they could have earned in normal markets. The Home Depot carries wood products that are certified as having been grown using sustainable forestry methods. Royal Dutch Shell screens all of its contractors and suppliers to ensure that none of them uses child labor. McDonald's has recently formed a partnership with the Center for Environmental Leadership to ensure that its suppliers use sustainable agriculture and conservation in growing their crops and raising their animals. Twenty-seven large U.S. corporations that buy more than $1 billion worth of paper, packaging, and pulp products each year have committed themselves not to purchase any such products that are harvested from old-growth forests.

At the heart of developing effective sustainability-based supplier relationships is conducting regular supplier audits. Shell uses such audits to monitor the child labor practices of its suppliers, and the apparel and textile manufacturers in the United States have developed a set of standards for working conditions, wages, and other terms of employment for their suppliers in developing countries. The Time, Inc. division of AOL Time Warner conducts audits of all of its paper suppliers to ensure that the paper in its magazines is produced, printed, and transported in the most environmentally responsible ways possible, and WalMart regularly audits the environmental practices of it suppliers.[32]

Interest Groups

For years, private citizens have brought significant social and environmental pressures to bear on business organizations through the activities of social and environmental interest groups. Three very prominent and very different examples of successful efforts to protect the natural environment by environmental interest groups are the Legal Defense Fund's efforts to protect the northern spotted owl, Environmental Defense's efforts to convince McDonald's to reduce solid wastes, and Greenpeace's campaign to stop Shell Oil UK from sinking the Brent Spar oil platform in the North Sea. The Legal Defense Fund used the courtroom as its platform to restrict logging in the old-growth forests of the northwestern United States on the grounds that such logging violates the Endangered Species Act. By contrast, Environmental Defense developed a cooperative alliance with McDonald's that eventually resulted in a major overhaul of the fast food giant's packaging and waste-management practices. Finally, Greenpeace's efforts against Shell were directly confrontational; Greenpeace activists physically boarded and chained themselves to the Brent Spar platform and refused to leave, and Greenpeace ships, loaded with both activists and journalists, followed and harassed the Shell tugboats as they tried to move Brent Spar 120 miles to the dump site.

Two influential interest groups committed to human rights and social justice are Amnesty International and the Southern Poverty Law Center. Amnesty International, founded in 1961, is dedicated to preventing grave human rights abuses—such as inhumane or unjust imprisonment, torture, political murders, and so forth—across the globe. There are 7,800 Amnesty International groups operating in more than 100 countries. Amnesty International attempts to directly influence the human rights practices of business organizations through its Business and Economic Relations Network, which encourages business organizations to write codes of conduct that explicitly recognize the value of human rights in all of their practices. Founded in 1971, the Southern Poverty Law Center (SPLC) in Montgomery, Alabama, conducts legal actions against White supremacist groups and other hate groups, sponsors educational programs for teaching tolerance in schools and business organizations, and has recently completed the Civil Rights Memorial, which commemorates U.S. civil rights activists who lost their lives so that others could be free to pursue their economic, educational, and social dreams without regard to their heritage. SPLC's founder, Morris Dees, has personally won several hate crime lawsuits across the United States; his court victories have bankrupted large hate groups like the United Klans of America and the Aryan Nations.[33]

These examples point to some significant differences regarding how interest groups attempt to influence the performance of business organizations, including cooperation, education, litigation, and confrontation. In fact, research indicates that interest groups differ along four continuums. First, they have different ecological philosophies, ranging from those with emotional, spiritually based philosophies to those with technological, scientific, rational, data-oriented philosophies. Second, interest groups differ in their methods of advocacy; at one end of this continuum are groups that are direct, confrontational, and persistent, and at the other end are groups that are indirect, low key, and willing to cooperate and collaborate. Third, interest groups differ according to their desired end states, that is, their beliefs about what segment or segments of society need changing, how much they need changing, and how urgent the need for change is. Finally, interest groups differ along structural dimensions, including professionalism, size, and complexity. Of these four differences, research indicates that the method of advocacy is the key dimension because it is indicative of the level of cooperation that will likely exist between the interest group and the organization. At the confrontational end of the method of advocacy dimension, the interest group and the organization will experience significant conflict and resistance to each other's demands. On the other hand, at the collaborative end, the interest group and the organization will experience cooperation and the opportunity to reach a true consensus and understanding. Regardless of the method of advocacy, however, managers should understand all of the diverse approaches of interest groups, and they should develop cooperative relationships with these groups based on a mutual search for sustainability if at all possible.[34]

Lenders and Insurers

Environmental problems bring with them the potential for legal liability, financial liability, property damage, and property loss. This means that environmental problems are concerns for the financial institutions that extend credit and insurance to business organizations. For this reason, lenders and insurers today regularly require environmental audits, and they are less willing to finance and insure environmentally risky projects, especially in the wake of court decisions that often make current owners responsible for environmental problems created by previous owners.

It goes without saying that lenders and insurers are stakeholders with a tremendous influence on organizations. Most business activities simply cannot survive without sufficient financial backing or insurance. Thus, business ventures today cannot really be considered feasible unless the potential financial liabilities related to property damage, personal injury, cleanup, and natural resource costs are carefully considered. Because of the problems created for property owners related to contaminated land, there has been a huge increase in the demand for environmental insurance. Formerly, only firms with obvious environmental risks, such as chemical and waste-management firms, demanded this type of insurance. However, this has changed; many small manufacturers and small retailers (such as dry cleaners and gas stations) are now requesting environmental insurance. Residential construction contractors are also demanding more environmental insurance, especially in areas where contaminated land is abundant and for sale (i.e., land on closed military bases). Without such coverage, restoring contaminated lands is usually too expensive, and the land may remain contaminated forever.[35]

One of the most encouraging social responsibility trends among lenders is the growth in community-based banking. Banks and other lenders in many communities across the globe are now making a concerted effort to keep a significant amount of the money they loan in the local community. They are making more loans to local small businesses, minorities, women, and low-to-moderate-income families. Such lending practices energize communities economically as well as socially by keeping the money working in the community, giving the community the benefit of the money's true multiplier effect. This trend is very much a part of community going-local efforts such as those discussed in Chapter 4.

Industry Standard Setters

A number of standard-setting initiatives focusing on improved environmental performance are coming from trade associations, quality associations, and other groups. In general, the approach of these groups is to establish environmental standards that, although voluntary, are required for group membership, certification, and so on. Some initiatives of this type that have had especially strong influences on industry environmental standards include the CERES Principles (discussed above in the section on Investors) and the European Union's Eco-Management and Audit Scheme (discussed above in the section on Regulators). Others include the Chemical Manufacturers Association's Responsible Care guidelines, the International Chamber of Commerce's Business Charter for Sustainable Development, the British Standards Institution's Standard 7750, and the International Standards Organization (ISO) 14000 standards.

In March 1990, the members of the Chemical Manufacturers Association (CMA) agreed to enact a set of guidelines designed to significantly improve the way firms in the industry managed the environmental aspects of the chemical manufacturing process. These Responsible Care guidelines are modeled after similar guidelines enacted by chemical manufacturers in Canada under the guidance of Dow Canada CEO David Buzzelli. There are nine Responsible Care guidelines: (1) To safely develop, produce, transport, use, and dispose of chemicals; (2) to make health, safety, and the environment priority considerations in planning for both current and new products; (3) to promptly report any chemical or health hazards and to be prepared to deal with them if they occur; (4) to inform customers how to safely transport, store, and use chemicals; (5) to always operate plants in a safe manner; (6) to support research on the environmental impacts of products, processes, and wastes; (7) to contribute significant efforts to resolve problems caused by past practices; (8) to participate with the government to develop laws and regulations that promote a safer, more environmentally sound industry; and (9) to share environmental management experiences and information with other firms in the industry.

The International Chamber of Commerce (ICC) adopted the Business Charter for Sustainable Development in November 1990. The charter was developed through a cooperative effort involving the United Nations Environmental Programme, the International Environmental Bureau, and the

Global Environmental Management Initiative (GEMI), along with the ICC. In the charter, the ICC outlines sixteen principles that encourage a wide range of enterprises to commit themselves to improving their environmental performance and social responsibility.[36]

The British Standards Institution's BS 7750 environmental management standards are quite comprehensive. BS 7750 is essentially an internal environmental management system that includes processes for establishing environmental objectives, implementing those objectives, and measuring and reporting on the firm's progress toward accomplishing those objectives. Although meeting the requirements of BS 7750 is technically voluntary, firms wishing to do business in the United Kingdom will certainly feel pressure from their customers and clients there to comply with it.

The BS 7750 standards served as models for the development of the ISO 14000 standards, which constitute the closest thing yet to a set of truly global environmental management standards. The ISO 9000 quality standards have always had an environmental component, but ISO 14000 focuses exclusively on environmental management. Like the ISO 9000 quality standards, ISO 14000 certification is a requirement in many European markets. The ISO 14000 standards are currently under review by the European Parliament, but significant revision is not expected.[37]

There are also numerous standard setting initiatives focusing on improved social performance. For example, the Council on Economic Priorities Accreditation Agency (CEPAA), an arm of the CEP (mentioned above in the section on Investors), has published a social accountability standard based on the United Nations Declaration of Human Rights and the conventions of the International Labour Organization (ILO).[38] These and a number of other environmental and social standard setters will be covered further in Chapters 6, 8, and 10.

Conclusions

In this chapter, we discussed internal strategic advantage analysis as it relates to sustainable strategic management. Specifically, we discussed one of the primary tools for internal analysis, value-chain analysis. To be used effectively as a tool for determining strategic advantages in sustainable strategic management, value-chain analysis needs expanding beyond its narrow supplier–creation–manufacture–consumer focus to include both nature and the greater social system. We also discussed the need to close the value chain, shifting it from a linear ecoefficiency and socioefficiency focus to a circular ecoeffectiveness and socioeffectiveness focus. Finally, we presented a discussion of the role of stakeholders in determining competitive advantages of the firm, and we suggested that there are significant numbers of stakeholders in the marketplace today to support firms pursuing sustainable strategic management, including regulators, consumers, investors, employees, suppliers, interest groups, lenders, insurers, and industry standard setters. We have now examined both external environmental analysis (Chapter 4) and internal strategic advantage analysis (this chapter) as they relate to sustainable strategic management. The question is, how are the results of these analyses molded and shaped into effective strategies for sustainability? We will respond to this question in the next chapter.

6

Sustainable Strategic Management Strategies

As discussed in Chapter 3, in sustainable strategic management a strategic vision based on the core value of sustainability serves as the foundation for formulating a firm's strategies. It also serves as a basis for determining what information the organization considers important and the ways the organization measures its success in terms of its economic, social, and ecological performance. Thus, a strategic vision based on sustainability shapes a firm's sustainable strategic management (SSM) strategies.

A strategy is a comprehensive plan or stream of decisions that relates the strategic advantages of a firm with its external opportunities and threats in order to accomplish organizational goals and objectives. Therefore, an analysis of the external environment (Chapter 4) and an assessment of the organization's strengths, core competencies, and weaknesses (Chapter 5) as they relate to economic, social, and environmental performance are necessary prerequisites for formulating SSM strategies. As we said in Chapter 3, we envision SSM strategies as integrative strategies designed to provide long-term competitive advantages to organizations by taking advantage of external opportunities and minimizing external threats along all three dimensions of sustainability. As discussed in Chapter 1, sustainable strategic management requires viewing the economy as an open living system rather than a closed circular flow. Thus, formulating SSM strategies requires systems thinking and generative learning. Regardless of whether SSM strategies are formulated via planned, emergent, incremental, or muddled-through processes, they must be based on an open systems planning model that represents the collective wisdom of the firm and its stakeholders with regard to its economic, social, and environmental performance.[1]

In essence then, SSM strategies are what organizations that "stand for sustainability" do. It is through these strategies that the philosophies and ethics of sustainable strategic management become tangible. SSM strategies are designed as vehicles for operationally integrating the ecosystem and the greater society into strategic decision-making processes. In short, SSM strategies provide valuable avenues for bringing the ecological, social, and economic dimensions of an organization's strategic vision to life. Like all business strategies, SSM strategies exist at the corporate, competitive (business), and functional levels of the firm. Thus, SSM strategies exist in a hierarchy with the transcendent core value of sustainability underlying each strategy level (see Figure 6.1). The hierarchy of SSM strategies will be the basis for the discussion in this chapter.

Functional Level Sustainable Strategic Management Strategies

Functional level strategies are designed to accomplish short-run objectives that in turn lead to the achievement of organizational goals. As discussed in Chapter 5, the identification of a firm's core competencies is the result of a strategic advantage analysis within the functional levels of the organization. Thus, functional level strategies are important in establishing and exploiting the core competencies of the firm. They give specific short-term guidance to operational managers in areas such as operations, finance, marketing, and human resources. Functional level strategies exist for every value chain activity and are essential for the implementation of competitive level SSM strategies (discussed in Part III).

An essential tenet of formulating SSM strategies is life-cycle analysis (LCA). LCA is a total systems approach designed to provide a cradle-to-grave appraisal of the ecological and social impacts of the firm's products and processes all along the type II value chain.[2] Not only does LCA provide strategic managers with a tool that can give them data on how to improve their environmental and social performance, it also provides a tool that can help them improve their competitive position in the marketplace. LCA is philosophically consistent with the formulation of functional level strategies that develop and exploit the core competencies that provide the competitive advantages afforded by competitive level SSM strategies. However, currently LCA is primarily focused on ecological impacts. Sustainable strategic management requires that this process be expanded to include the social impacts of the firm's products, services, and processes as well as be based on the cradle-to-cradle, type III closed-loop value chain.

Currently, most functional level strategies focus on ecoefficient environmental concerns. Called strategic environmental management (SEM) strategies, they vary in content depending on their location within the type II value

Figure 6.1 **Hierarchy of Sustainable Strategic Management Strategies**

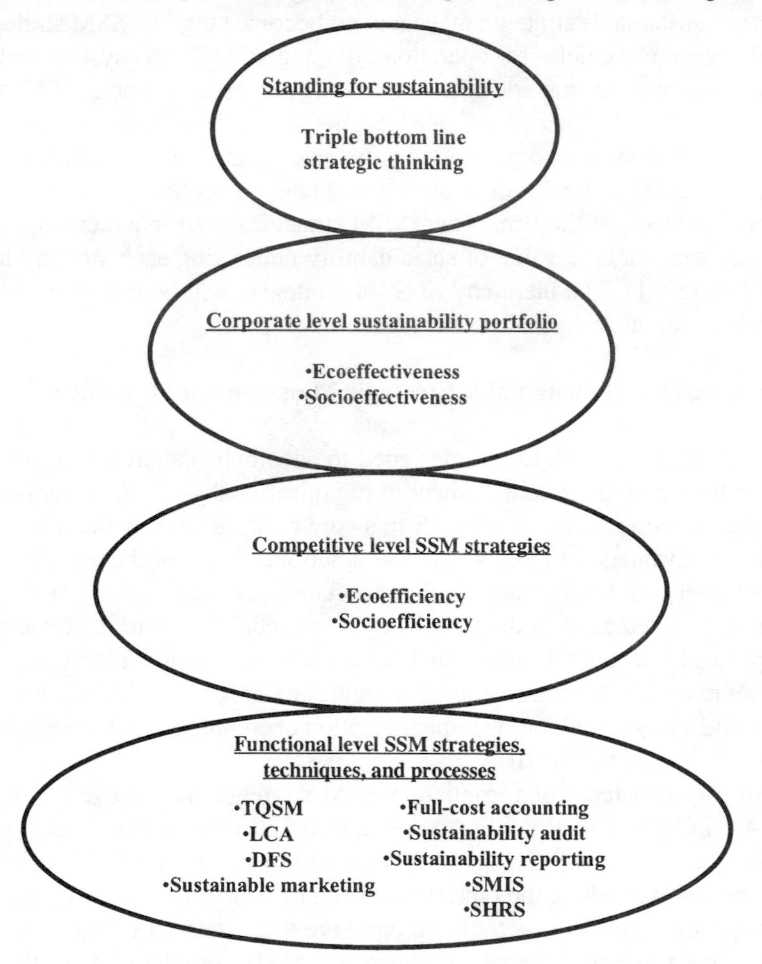

chain. Among the functional level SEM strategies currently associated with the primary activities of the type II value chain are design for environment (DFE), total quality environmental management (TQEM), and environmental marketing. Sustainable strategic management will require that such functional level SEM strategies be expanded in scope in the future to include the firm's social performance as well as its environmental performance, thus becoming SSM strategies that are based on the type III closed-loop value chain. Thus, DFE may become design for sustainability (DFS), and TQEM may become total quality sustainability management (TQSM). Many corporations have already expanded their environmental marketing efforts to include social cause marketing, and this trend is continuing.[3]

The functional level SEM strategies associated with the support activities of the type II value chain include full cost accounting, environmental auditing, environmental reporting, environmental management information systems (EMIS), and human resource management systems. As with SEM functional level strategies related to primary value chain activities, organizations will need to expand the scope of their SEM support activities in the future to include social performance as well as environmental performance. That is, like their primary activity counterparts, SSM support activity strategies are based on the type III value chain. Thus, full cost accounting will need to account for social as well as environmental costs, and environmental audits will need to be expanded to include both social and environmental performance (a sustainability audit). Environmental management information systems will be expanded to include social information as well as environmental information, becoming sustainable management information systems (SMIS). Human resource management systems will be designed to reinforce and support a sustainability culture within the organization by including social and environmental performance in the measurement, reporting, appraisal, and reward systems becoming sustainable human resource systems (SHRS). Many firms have already expanded the scope of their environmental audits to include their social as well as environmental performance. According to KPMG, 45 percent of the largest 350 companies now issue sustainability reports.[4]

The effective formulation and implementation of functional level SSM strategies are critical for developing and exploiting the core competencies on which the firm can build its competitive strategies necessary to achieve its sustainability goals. (Note: These functional strategies will be examined in depth in Chapters 8, 9, and 10.)

Competitive Level Sustainable Strategic Management Strategies

Competitive level strategies focus on how to effectively compete in an industry or a particular product/market segment. These strategies are based on identifying customer groups and needs and effectively segmenting the market to meet those needs. It was during the 1990s that many strategic managers recognized that managing ecological issues was critical for their organizations' competitive positions. Market opportunities for environmentally sensitive products and services appeared, and SEM emerged as a concept that provided strategic managers with the skills, processes, and functional level strategies for managing ecological issues while simultaneously creating value for the firm.[5] SEM strategies are based on the type II value chain model, they directly link the organization's economic sustainability with environmental sustainability, and they can be classified according to the nature of the competitive advantages they provide, that is, advantages based on lowering costs or providing opportunities for market differentiation.[6]

Strategic Environmental Management Cost Leadership Strategies

SEM strategies that are designed to provide firms with cost advantages through improved environmental efficiency are generally referred to as pollution prevention strategies.[7] These strategies are designed to improve environmental performance during the manufacturing process, thus preventing pollution before it happens. The pioneer in pollution prevention strategies was 3M. In 1975, 3M (at the time, one of the world's largest polluters and waste generators) began its Pollution Prevention Pays (3P) program. The program was designed to switch the focus of pollution management at 3M from the output end to the input end of the firm's production processes. In the first seventeen years of the 3P program, there were some 3,500 3P projects carried out at 3M, and the firm reported saving about $550 million as a result. These savings came from an annual reduction of 575,000 tons in air pollution, water pollution, sludge, and solid wastes as well as a reduction of 1.6 billion gallons per year in wastewater generation.

Core competencies in operations and logistics are the basis for the competitive advantages that are exploited via pollution prevention strategies. At the heart of these strategies is ecoefficiency (as mentioned in Chapter 5), which is creating value for business by doing more with less within the type II linear value chain. There are at least four dimensions of ecoefficiency that make it an important tool in developing competitive advantages: (1) dematerialization, where knowledge flows are substituted for material flows and products are designed with less resources and wastes, (2) closed-loop production systems and zero-waste manufacturing facilities, (3) service extension, where customized responses to consumer demand offer consumers the choice of leasing goods rather than buying them outright, and (4) functional extension, where smarter products with enhanced functionality and durability are being developed.[8]

Ecologically, strategies based on ecoefficiency serve to reduce resource depletion, materials use, energy consumption, emissions, and effluents while lowering costs. Ecoefficiency techniques often include redesigning pollution and waste control systems, redesigning production processes to be more environmentally sensitive, using recycled materials from production processes and outside sources, using renewable energy sources, increasing the durability of products, and increasing the service intensity of goods and services. Innovation is the strategic driver in improving ecoefficiency, and there are now significant data that demonstrate a direct relationship between ecoefficiency, competitiveness, and financial performance.[9] Ecoefficiency techniques provide the means for firms to enhance their economic and environmental performance via cost savings by reducing energy consumption, waste, and pollution while increasing recycling. Because of demands for external scrutiny of environmental operations, ecoefficient pollution prevention strategies not only provide firms with competitive advantages via cost reductions, they also provide firms with opportunities to establish social legitimacy in the greater community.

Pollution prevention has reaped big rewards for many organizations. Xerox Corporation, for example, is known for its remanufacturing, reuse, and recycling efforts. The company credits its product take-back efforts as well as other ecoefficiency techniques with producing $250 million in annual savings. Ninety percent of Xerox-designed equipment is remanufactured. Not only have these strategies improved Xerox's economic performance, they have also helped recruit and retain talented employees. Customers and other stakeholders have also been impressed with the numerous environmental awards that Xerox has won, thus enhancing its reputation and social legitimacy in the larger community.[10]

Strategic Environmental Management Differentiation Strategies

The second classification of SEM strategies consists of market-driven, product stewardship strategies.[11] Economically, the market-driven component of these strategies is designed to provide firms with competitive advantages by allowing them to ecologically differentiate their products and services from their competitors in the marketplace. Ecologically, these strategies reflect product stewardship, the idea that environmental hazards and life-cycle costs should be minimized in products or services. Firms moving from pollution prevention to product stewardship expand from a focus on materials, energy, and process efficiency to a focus on the complete life-cycle impacts of their products and processes. Product stewardship strategies are market-driven in that they provide sustained opportunities for firms to differentiate themselves from their competitors; but they go well beyond this. Basing product stewardship strategies on life-cycle analysis means that this differentiation is achieved by attending to both process and market factors in strategy formulation and implementation. Further, total product stewardship is best achieved by involving all of the firm's external stakeholders—suppliers, consumers, environmentalists, regulators, the community, and so forth—in the product development processes.

Marketing and product/service development are the functional levels at which core competencies must exist in order to create environmental differentiation. Some of the activities that can comprise this classification of strategies include entering new environmental markets or market segments, introducing new environmentally oriented products, redesigning products to be more environmentally sensitive, advertising the environmental benefits of products, redesigning product packaging, selling byproducts once discarded as wastes, and so forth.

Environmentally differentiating products and services can lead to increased revenue, market share, and profits.[12] Differentiation may result from increased brand value and brand equity, firm reputation, or customer attraction. Effective environmental differentiation is achieved through competitive product positioning in carefully segmented markets. Ecolabeling enhances environmental differentiation leading to potentially higher prices and expanded market share as well as satisfying consumer concerns about the environmental and social impacts of the products they purchase.[13] Ecolabeling is a marketing strategy that is currently being used worldwide to inform consumers about products from construction materials and household appliances to paints and paper products. The German Blue Angel ecolabeling system, which began in 1978, now covers 4,885 products in seventy-nine product groups from 944 manufacturers. One of the better known examples of ecolabeling entails the standards established by the Forestry Stewardship Council (FSC), an independent certification system, to evaluate the performance of wood products against a set of environmental, social, and economic standards. More than 600 companies, including IKEA and Home Depot, have joined trade networks committed to buying FSC-certified wood. Fish and organic food producers are also differentiating their product offerings via ecolabeling strategies.

A key to effective environmental differentiation is to focus marketing efforts on the market segment most prone to buying environmentally sensitive goods and services. This market segment is currently comprised of consumers that are better educated and wealthier than the general population. This is a small market niche, consisting of 10 percent to 12 percent of the population.[14] However, data indicate that this market segment is expanding rapidly.[15] For example, the $20 billion retail market for certified organic foods is growing between 10 percent and 30 percent annually in industrialized nations. More than 130 countries produce certified organic food for the retail market, including at least 65 developing countries. Water scarcity and sanitation also provide opportunities for environmental differentiation. There is an estimated explosion in the demand for water purification systems, and ITT and GE have already positioned themselves in this niche. The demand for water-efficient processes or products is also expected to grow significantly. Thus, a key to successful environmental differentiation is the competitive positioning of products and services in carefully segmented markets. To do this effectively requires being innovative, thus creating first-mover advantages. Continuous innovation is essential to maintaining a competitive advantage because as soon as competitors can imitate environmentally sensitive offerings, competitive differentiation is lost.

Electrolux is one firm that has been successful in environmentally differentiating its product line. Electrolux discovered that more than 90 percent of the environmental impacts of its appliances occurred after consumers purchased them. This was the impetus for Electrolux engineers to design an environmentally improved product line consisting of solar-powered lawn mowers, chain saws lubricated with vegetable oil, high efficiency household appliances, and so forth. The company found that consumers were willing to pay a premium price for the long-term environmental benefits and operating savings, so it expanded its product offerings to include other environmentally sensitive appliances, such as water efficient dishwashers and washing machines. In 1996, Electrolux found that its environmentally sensitive products returned a profit margin 3.8 percent higher than its traditional products, and by 1998, revenues from what it calls its Green Range line of appliances accounted for 16 percent of total sales and 24 percent of gross margins.[16]

Strategic Environmental Management to Sustainable Strategic Management Competitive Level Strategies

Although SEM strategies have effectively integrated economic and environmental performance, they have not yet fully incorporated the firm's social performance. As mentioned in Chapter 5, socioefficiency is the ability to enhance both the social capital of the community and the human capital of the firm while contributing to the organization's economic sustainability.[17] SSM competitive level strategies expand SEM strategies to include the social performance of the firm, integrating the firm's corporate social responsibility (CSR) with its economic performance. According to Charles Holliday, CEO and chairman of DuPont, Philip Watts, chairman of the managing directors of Royal Dutch Shell, and their coauthor Stephen Schmidheiny, "Corporate officials cannot maximize returns to shareholders over the long term without engaging in, and being seen to engage in, CSR."[18] These authors also believe that firms will realize their investment in CSR only when that investment is focused on and connected to its core business strategy.

Investment in the human capital of the firm can be both socially responsible and economically profitable.[19] Research by McKinsey Consulting Company indicates that the most important corporate resource over the next twenty years will be talented employees who are smart, technologically skilled, globally astute, and operationally agile. They found that the battle for talented employees is dramatically intensifying, and they concluded that attracting and retaining talent is a business imperative for economic survival. Firms with progressive human resource policies and strategies based on an inspiring vision like sustainability have been shown to have a competitive edge in attracting and retaining high quality employees. Strategic managers who invest in the human capital of their organizations do so because they believe their investment is a means to greater productivity and profitability as well as a means of fulfilling their social responsibilities. For example, Johnson and Johnson has decentralized power and decision making down to business units, allowing employees to have more freedom to question everything they do. CEO Ralph Larsen believes that this questioning culture makes Johnson and Johnson more innovative and thus more economically successful. Other firms have found that supporting employee volunteerism has a direct influence on employee commitment and retention. Thus, developing and utilizing employee capital and knowledge leads to motivation, commitment, innovation, and higher performance levels.

Strategic managers have long recognized that regulatory compliance, philanthropic contributions, and investments in the communities where they operate can lead to an improved public image, social legitimacy, increased credibility, and consumer trust. These results generally translate into improved economic performance. For example, Racine, Wisconsin, is the headquarters for 166-year-old SC Johnson Enterprises. Racine, like many mid-sized industrial cities in the United States, has experienced rising unemployment because of a lack of skilled labor, increasing crime rates, and poorly performing public schools. During the mid-1990s the local economy was hit hard, new firms found locating there less attractive, factories were being vacated, leaving desolate brownfields, and the local tax base was eroding. In response to these problems, SC Johnson became the catalyst for Sustainable Racine, a local community partnership committed to making Racine a better place to live. Today, local teachers and parents are working together for better student performance and teacher training, commercial vacancy rates in downtown Racine have fallen from 46 percent to 18 percent, and four municipalities are working together to share water and sewer costs. SC Johnson is also committed to helping other communities in which it operates worldwide. For example, in South Africa it provides a teacher for HIV-positive orphans and daily meals for 300 children, and it operates an adult education center where over 1,000 local women have been trained. Efforts like these make the communities in which SC Johnson operates more viable and sustainable while affording the company a more skilled, productive workforce and a reputation for being a good corporate citizen. These efforts translate into enhanced human and social capital as well as increased economic performance for the firm. SC Johnson's strategic managers are walking their talk by instrumentally living their core value of "every place should be a better place because we are there."[20]

Thus, a firm's reputation as a good corporate citizen can be used to create a perceived differentiation in its products and services. Social labeling has recently emerged as a means for firms to achieve this. Social labeling provides consumers with assurance about the social and ethical impacts of the organization's processes and products. The Ethical Trade Initiative, the Fair Trade Foundation, and the Clean Clothes Campaign all aim to assure consumers that the conditions within a firm's production chain meet basic standards and do not utilize child labor, bonded labor, or sweatshops in the production of their products. Social labeling criteria are usually based on the conventions of the International Labour Organization. Reebok, for example, found that incorporating these internationally recognized human rights standards into its business practices has led to improved worker morale, a better working environment, and higher quality products. A number of companies, including BP and Shell, have incorporated elements of the United Nations Declaration of Human Rights into their competitive strategies. Regardless of the means, having corporate social responsibility connected to the core of a firm's competitive strategy reaps benefits for employees, local communities, and shareholders.[21]

In sum, strategic managers employing competitive level SSM strategies recognize that ecological and social sustainability are intricately tied to the economic sustainability of their firms, recognizing the importance of ecological and social performance in creating shareholder value. Pollution prevention, product stewardship, and community-based CSR are adaptive strategies that organizations can employ to do less harm to the environment, benefit their local communities, and allow firms to reap substantial economic benefits. Many suggest that the consistently positive economic returns related to competitive level SSM strategies result from the fact that these strategies are designed to pick the low-hanging fruit—that is, they are designed to take advantage of those social and environmental opportunities with the greatest chance of short-term payoffs. However, as will be demonstrated in the next section, these short-term economic benefits will not be nearly as available or easily reaped for corporate level SSM strategies.

The Sustainable Strategic Management Corporate Portfolio

Corporate strategy is the overall plan for a diversified organization. The focus of corporate strategy is on managing the mix, scope, and emphasis of a firm's portfolio of strategic business units (SBUs), exploiting the synergies among its lines of businesses, and deploying resources accordingly. Corporate strategy is the way a corporation seeks to create value through the configuration of its SBUs and the coordination of its multimarket activities. Via an effective corporate level strategy, a firm's portfolio of businesses should create more value together than it would if each SBU stood alone.

SSM corporate level strategies are generative strategies designed to manage the configuration of the firm's portfolio of SBUs and processes in ways that create synergy among the firm's economic, social, and ecological performance along the type III closed-loop value chain. Thus, we envision an SSM corporate portfolio as a set of integrative processes that facilitate managing the various SBUs that are implementing SSM competitive level strategies. Whereas the strategies of each SBU are typically focused on developing and exploiting core competencies related to the business unit's economic, social, and ecological performance in product/market segments, the SSM corporate portfolio focuses on the processes of generative learning, entrepreneurial learning, dialogue, and organizational change processes necessary to move toward sustainable strategic management. These processes provide the vehicles for strategic managers to question the underlying assumptions of their corporate portfolios, to develop innovative approaches for sustainable product and service introductions, and to engage network partners and other stakeholder groups. Thus, the SSM corporate portfolio should provide more progress toward sustainability than one SBU operating alone.

As previously discussed, corporate level sustainable strategic management requires transformation to a view of organizations as parts of a living system (see Figure 1.2 and Figure 5.3). This mental model provides the appropriate foundation for sustainable strategic thinking and decision making. Within this framework, strategic managers view their employees as human capital with intrinsic as well as instrumental worth. Also, within this model, the corporate level strategic managers operate at the SSM intelligence level, integrating spiritual, intellectual, and emotional intelligence (see Chapter 3). They are able and willing to question the fundamental assumptions underlying their portfolios, pondering "what can be." By operating at the SSM intelligence level, CEOs are able to integrate the complex issues of how to care for future generations, how to care for their organizations, and how to care for the earth's stakeholders.

Strategic managers define the corporate mission of the firm by analyzing the purpose, scope, and balance of the firm's portfolio. Defining the SSM portfolio's mission requires the same processes mentioned above—generative and entrepreneurial learning, dialogue, and organizational change management. These allow CEOs to define the purpose, scope, and balance of the firm's portfolio in terms of sustainability, thus achieving both a balance and synergy among the firm's economic, environmental, and social pursuits.

The purpose of the SSM corporate portfolio is to provide a framework that allows sustainable strategic managers to continually question the fundamental assumptions of their portfolios in light of both eco- and socio-effectiveness. This generally shifts the firm's focus to global sustainability where the link between social, ecological, and economic performance is not always as direct as it is at the business unit level where eco- and socioefficiency provide more direct ties among these three. Using eco- and socioefficiency as a basis, strategic managers at the business unit level calculate the relative social and ecological costs and benefits before formulating competitive level SSM strategies. However, SSM corporate level strategies should address the absolute limits of natural and social capital because of the irreversibility, nonlinearity, and nonsubstitutability that often characterize precious social and natural capital. An effective SSM corporate portfolio should contribute to the preservation of both social and natural capital by having absolute, positive social and ecological impacts. Thus, the purpose of the SSM corporate portfolio is to make the world a better home for current and future generations by making positive contributions to its sustainability.

To fulfill this purpose, firms will have to expand the scope of their corporate portfolios to include issues such as poverty and income distribution in the developing nations. This expands the portfolio from a focus on local community sustainability to the broader focus of global sustainability, addressing the gaps between developed and developing countries of the world. The scope of the portfolio will also have to expand to address the issues of overconsumption and waste in the developed world, where 80 percent of the approximately $20 trillion in annual worldwide private consumption occurs.[22] Therefore, sustainable consumption, consumption limited by the carrying capacity of the planet, should be a critical strategic driver for corporate level SSM strategies. Thus, whereas SSM competitive strategies are directly linked to economic performance, enabling the low-hanging social and environmental fruit to be profitably picked, the expanded scope of the SSM corporate portfolio includes concern for income, consumption, and waste gaps between the developed and developing worlds, requiring strategic managers to adopt posterity as their appropriate planning horizon. Thus, payback for many investments is not likely to be realized in the short run because there is relatively little low-hanging fruit ripe for profitable plucking in these emerging markets.

Balancing the corporate portfolio traditionally refers to balancing the cash flow among the firm's various SBUs. However, balancing the SSM corporate portfolio refers to balancing the cash flow from the traditional product and service SBUs with the cash flow from business units created to address the opportunities arising from the expanded social and ecological scope and purpose of the portfolio. Thus, SSM corporate level strategies are designed to take advantage of new market opportunities in the developing world while contributing to the social and ecological sustainability in these emerging markets. SSM corporate strategies are also designed to address the issues of waste and overconsumption in the developed world by delivering value to consumers while minimizing the throughput of energy and matter. These strategies should eventually focus on changing consumption patterns in order to preserve natural and social capital. Therefore, the SSM corporate portfolio (and thus the firm's corporate mission) has the unique purpose of making positive contributions to a sustainable world by expanding its scope to meet the needs of people in the developing world and addressing the issues of sustainable consumption in the developed world, while ensuring that its SBUs are balanced so that the portfolio contributes to the long-term economic sustainability of the firm.

Corporate Level Sustainable Strategic Management Strategies

The developing world is full of emerging markets providing long-term opportunities for firms engaged in sustainable strategic management. However, taking advantage of these long-term economic opportunities requires that strategic managers engage in generative, entrepreneurial learning that will allow their organizations to develop products and services that meet the needs of the 4 billion aspiring but economically deprived people who reside in these markets.[23] C.K. Prahalad and Stuart Hart believe that the world's poorest people, those with annual per capita income less than $1,500 (what they call the "bottom of the pyramid"), afford multinationals economic opportunities as well as significant managerial challenges as they attempt to develop and sell products and services that are culturally sensitive, ecologically sustainable, and profitable. These markets often do not have the infrastructure or products even to meet the basic needs of their disadvantaged populations. Thus, they provide excellent potential markets for new environmentally and socially sensitive technologies, products, and services designed to raise the poor up from their current poverty, social and environmental decay, and political chaos. Serving these markets, however, requires changing perceptions about the developing world. One common perception that needs to be changed is that the poor cannot afford or cannot use products sold in the developed world, and another is that managers will not be willing to work in these markets. Prahalad and Hart contend that taking advantage of market opportunities in developing nations will require strategic managers to think differently about how to develop and market products that are low cost, good quality, sustainable, and profitable. Those that can achieve this, according to these authors, can reap significant financial gains at the bottom of the pyramid. However, others disagree, citing the specter of long payback periods for investments in emerging markets.[24] Regardless, both the purpose and expanded scope of the SSM corporate portfolio encourage participation in emerging markets as part of a viable economically, socially, and ecologically balanced portfolio.

As mentioned above, market opportunities in emerging markets require strategic managers to think differently and to develop innovative products and approaches to the marketplace. Recall we discussed in Chapter 4 that the growing population of young people in the developing world provides opportunities for labor and new consumer markets as the developed world's population ages.[25] The food system in the developing world neither meets the people's needs nor is it environmentally sustainable. Opportunities exist for the numerous industries that are involved in the food supply chain to create new products and processes to sustainably grow, process, and deliver affordable food to those markets where people are experiencing severe malnourishment. Opportunities exist for firms to develop solutions to the many public health issues of the developing world, including pharmaceuticals, health services, water infrastructure, and sanitation, among others. Energy use is necessary to meet basic human needs, and opportunities exist for new markets in alternative energies, energy conservation, and energy-efficient technologies. Information technology affords numerous opportunities to facilitate the transition from resource-based to knowledge-based economies in the developing world. Innovative uses of information technology can also connect the aspiring poor to educational and entrepreneurial opportunities while providing market opportunities for firms willing to invest in these markets. By exploiting these long-run market opportunities, strategic managers can grow their firms while making positive contributions to the social and natural capital in the developing world.

Some firms have already developed innovative strategies to capitalize on these long-run opportunities in emerging markets.[26] For example, Unilever has stated that one of its strategic priorities is to penetrate the markets of the world's 4 billion poor people. To accomplish this, Unilever's Indian subsidiary, Hindustan Lever Ltd. (HLL), has developed detergent products for poor consumers who live in mostly rural areas and who wash clothes in rivers and streams. The detergent has a lower ratio of oil to water, making it less polluting to the public water systems. Via innovative manufacturing, pricing, distribution, and packaging strategies, such as single-serve packaging for poor consumers who cannot afford to buy in bulk, HLL has obtained a 38 percent share of India's detergent market. HLL has also developed innovative technology that provides practical, inexpensive, low energy refrigeration in India. The new technology allows major reductions in energy use, reduces the amount of polluting refrigerants, and is cheaper to build and to use.

Arvind Mills is another firm that has designed a sustainable product for the developing world.[27] It sells ready-to-make Ruf and Tuf Jean Kits for $6 each to a network of 4,000 tailors in India. These kits contain the ready-to-assemble components of denim, zipper, rivets, and a patch. By designing its manufacturing and distribution systems to take advantage of India's most abundant resource, labor, Ruf and Tuf has captured the largest market share of jeans in India, easily surpassing Levi Strauss and other brands (whose finished jeans cost between $40 and $60). By capitalizing on India's abundant labor, the firm is generating income while stimulating the local village economies.

Providing access to credit and increasing the earning potential of the poor are also crucial for improving social capital in emerging markets. The Grameen Bank Ltd. in Bangladesh pioneered a lending service for the poor more than twenty years ago. By effectively addressing the issues related to extending credit to the lowest income consumers, such as high credit risks, lack of collateral, and so forth, Grameen Bank now provides microcredit services in more than 40,000 villages in Bangladesh. In 1996,Grameen Bank achieved a 95 percent repayment rate from its 2.3 million customers, 95 percent of whom were women. Citigroup has also made a commitment to provide microloans of less than $15, and is experimenting in Bangalore, India, with 24-hour 7-day services for customers who have as little as $25 on deposit. In addition, in partnership with the United Nations, Citigroup has adopted a goal to provide basic credit to the 100 million poorest families in the world.

Bridging the digital divide between the rich and poor nations of the world also provides long-run market opportunities, leading to what some believe to be a digital dividend.[28] In today's electronic world, poverty of information represents a huge obstacle to sustainability. As mentioned in Chapter 4, more than half of humanity has yet to make a phone call, only 7 percent have access to a personal computer, and only 4 percent have access to the Internet.[29] Hewlett-Packard (HP) is attempting to bridge the digital divide with a stated vision of "world e-inclusion" that focuses on providing appropriate technology, products, and processes for the world's poor. Its new venture, World e-Inclusion Services, is designed to address several needs of this market, including health and telemedicine, education, broadened market access, increased employment, and expanded credit access. As part of this strategy, HP has entered into an alliance with MIT Media Lab and the Foundation for Sustainable Development in Costa Rica to develop and implement "telecenters" for remote villages that provide modern information technology equipment with high-speed Internet connections at an affordable price. Cisco Systems, in partnership with the United Nations, funds the Netaid.org Web site that supports their eAction program designed to fight extreme poverty by enabling people to make cash donations, volunteer their time, and buy arts, crafts, and foods.[30]

These are just a few examples of organizations that have looked to the developing world and asked themselves, "What can be?" Taking advantage of these market opportunities requires the radical rethinking of market dynamics. It takes tremendous creativity and entrepreneurship to engineer a market infrastructure out of a previously unorganized sector. Prahalad and Hart believe that the key to thriving in these markets is to create buying power, shape aspirations, improve access, and tailor local solutions.[31] Corporate level strategic managers willing to engage in generative and entrepreneurial learning, in dialogue, and in change management processes have the potential to participate in these emerging markets in ways that allow them to make positive contributions to global social and natural capital while developing long-run market opportunities.

As mentioned above, SSM corporate strategies also need to address the issue of sustainable consumption in the developed world. Sustainable consumption is consumption that is in balance with the carrying capacity of the earth. As noted in Chapter 2, the 25 percent of the earth's population that lives in the developed world controls most of the world's financial resources, consumes most of its goods and services, generates most of its wastes, and uses most of the natural resources extracted from the planet. Given that the remaining 75 percent of the people on the planet appear to want to emulate these consumption patterns, the environmental risks of rising consumption create major roadblocks to a more sustainable world. The money spent on household consumption worldwide increased 68 percent between 1980 and 1998, with most of this increase occurring in the developed world.[32] Given these rising consumption patterns, there are certainly abundant opportunities for organizations to develop innovative strategies that can lower resource intensity, reduce waste, and increase the durability of the goods being consumed, thus making a positive contribution to natural capital.

Sustainable consumption is not just about consuming less; it is about changing the form of consumption.[33] Sustainable strategic managers must rethink how to create value for the consumer market, where consumers still demand quality, convenience, and content. Through innovative strategies aimed at changing consumption patterns, firms can create value for consumers while minimizing the environmental impact resulting from the production and consumption of goods. "In the new model, value is delivered as a flow of services—providing illumination, for example, rather than selling light bulbs."[34] This represents a shift in thinking from producing products, increasing throughput, and making energy and capital-intensive investments, to bundling services, selling end-use value, and ensuring cradle-to-cradle product stewardship. Interface's Evergreen Lease program is an example of this. Under this program Interface leases rather than sells clean, fresh carpet to customers for a monthly fee. Because 80 percent of the carpet wear takes place on 20 percent of the carpet, this results in a fivefold savings on carpet replacement. In addition, Interface has designed its carpet to last four times longer and to use 40 percent less materials than traditional carpet. These two combined factors have reduced resource intensity 35 percent while still providing customers with a superior floor-covering service. Dow Chemical and Safety-Kleen lease dissolving services rather than sell solvents because they can reuse the same solvent, thereby reducing costs. United Technologies' Carrier Division, the world's largest manufacturer of air conditioners, has changed its mission from selling air conditioners to leasing comfort, and Schindler, the elevator manufacturer, now sells vertical transportation services rather than elevators.[35]

A variation on the leasing model is called shared savings.[36] Currently, many chemical companies sell chemical performance rather than chemicals. Performance expectations are set and performance fees are paid each month. For example, Ford Motor Company has subcontracted with PPG/Chemfil to manage their Total Fluids and Total Solvent management programs. Ford pays a performance fee each month to PPG/Chemfil to manage the inventory, tracking, distribution, and environmental control, and quality control of the chemicals. The chemicals remain the property of PPG/Chemfil until Ford actually uses them.

Rethinking how to create value for consumers via selling services or leasing goods reverses today's throwaway society mentality. Rather than thinking in terms of planned obsolescence, manufacturers are motivated to make their products more durable and longer lasting. Profitability is a result of lowering the resource intensity of products and making products that last longer. In terms of environmental sustainability, this new approach to the marketplace provides for automatic product take back, thus ensuring cradle-to-cradle product stewardship designed to conserve natural capital.

In sum, managing at the corporate level in sustainable strategic management requires strategic managers to rethink their current paradigms, shifting the purpose of their portfolios of businesses to ones that stand for sustainability. If the core value of sustainability is truly at the heart of the corporate strategy formulation process, the scope of the firm's portfolio of businesses will expand to include strategies for making an absolute positive contribution to the social and natural capital in both the developed and developing worlds. This will require transformational change on the part of strategic managers and their organizations.

Conclusions

The distinguishing characteristic of SSM strategies is that they are formulated with the core value of sustainability underlying the formulation process. As mentioned previously, SSM strategies are what organizations that "stand for sustainability" do. They are the action components of sustainable strategic management, serving as the vehicles for achieving sustainability goals. The SSM hierarchy of strategies served as the organizing framework for this chapter, which covered the content of functional, competitive, and corporate SSM strategies.

As discussed, functional level SSM strategies exist for each value chain activity (primary and supporting) and provide the foundation for building core competencies in eco- and socioefficiency. SSM competitive level strategies exploit these core competencies in carefully chosen product-market segments via cost or differentiation competitive advantages. At the competitive business unit level, SSM strategies are adaptive strategies that provide a direct link between economic, ecological, and social performance, picking the so-called low-hanging fruit. However, these short-term benefits will not be nearly as abundant at the corporate level of SSM. At the corporate level, strategic managers must rethink their current business models and the underlying values of their corporate portfolios, building their strategies on eco- and socioeffectiveness. This requires adopting the type III value chain model as a basis for strategy formulation. In other words, managers must be committed to contribute, in an absolute sense, to the social and natural capital of the planet. This will require adopting posterity as their planning horizon. This allows them to look at their portfolios and ask themselves, "What can be?"

7

Sustainable Strategic Analysis and Choice

Strategic management scholars generally agree that a successful strategy must be formulated to take advantage of environmental opportunities and minimize the impact of environmental threats. A strategy must also be built on the firm's core competencies that are identified through the internal audit. Thus, an organization must achieve a strategic fit between its external and internal environments in order to survive and be successful. The most critical factor in the success of any organization is the quality of strategic decisions that top managers make in attempting to achieve this strategic fit.

The strategic decision-making process entails three sequential stages. The first stage is a data-gathering phase, which involves extensive research and analysis of the external and internal environments of the organization in order to provide sufficient and accurate inputs into the strategic decision-making process. The data gathered are then analyzed, and opportunities (O), threats (T), strengths (S), and weaknesses (W) are identified. The second stage in strategic decision making entails matching or fitting the external opportunities and threats with the strengths and weaknesses. Many techniques have been developed by management scholars to aid in this stage of strategic decision making. Most of these techniques consist of some type of matrix that allows strategic managers to match opportunities and threats with strengths and weaknesses. Matrixes such as TOWS and SWOT are utilized in developing feasible strategic alternatives that match key external and internal factors. Tables 7.1 and 7.2 demonstrate the strategic alternatives available to Anthony Flaccavento of Appalachian Sustainable Development and Bill Ford of Ford Motor Company (introduced in Chapter 1 and covered more thoroughly in Chapter 11), utilizing the TOWS matrix. Product portfolio matrixes such as the BCG, McKinsey, and GE, as well as the Grand Strategy matrix, also aid in the strategic decision-making process at the corporate level where strategic managers must manage the mix, scope, and emphasis of the corporate portfolio.[1]

The third stage of strategic decision making is one of strategic choice. Strategic choice involves evaluating the alternative strategies generated from the above analysis and deciding which set of strategies the firm should pursue in order to achieve its goals and objectives. At this stage strategic managers analyze the quantitative data that have been generated. However, strategic choice is not influenced only by quantitative analysis. It also entails mental processes such as judgment and intuition. Strategic decisions are influenced by top managers' perceptions of the degree of risk of the alternative strategies, their awareness of the success and failure of past strategies, and their perceptions of dependence on various stakeholder groups. Thus, strategic decisions are greatly influenced by strategic managers' perceptions of objective reality. Because perceptions are formed via cognitive processes, it is extremely important to understand what goes on in that black box called the mind when strategic managers make decisions that are critical to the organization's success.

Perceptions, Values, Assumptions, and Strategic Decision Making

There was a dramatic paradigm shift in science during the twentieth century. Theories of relativity and quantum physics led scientists to understand that there is no such thing as objective reality; there are only perceptions of reality. This new scientific paradigm has crept into management thought over the past two decades. Cognitive explanations of human behavior that focus on the relationships between what people perceive and what they do have begun to revolutionize the way many people think about how decisions in organizations are made.

Perceptions and Organizational Behavior

During much of the twentieth century, theories of why people behave the way they do have focused primarily on the external environment. These approaches advocate that behavior is entirely shaped by the influence of environmental stimuli and reinforcement. Research on these models has been quite fruitful. However, even in the face of considerable evidence, scholars and practitioners alike have resisted the idea that human beings are primarily motivated by rewards and punishments, saying that such theories leave little or no room for the influence of human cognitive processes

Table 7.1 **TOWS Matrix for Appalachian Sustainable Development**

	STRENGTHS (S) 1. Committed, experienced leadership 2. Highly qualified professional staff 3. Meaningful work 4. Solar wood-drying technology 5. Organic food-processing technology	**WEAKNESSES (W)** 1. Lack of marketing expertise 2. Lack of clear strategic/marketing plan 3. Not financially sustainable without outside funding 4. Low pay, high turnover among professional staff 5. Inconsistency of solar technology 6. Transportation capacity
OPPORTUNITIES (O) 1. Growing demand for organic food 2. Growing demand for sustainable wood products 3. Undeveloped local organic food market 4. Undeveloped local sustainable woods market 5. Untapped food service market 6. New sustainable energy technology	**S/O STRATEGIES** 1. Sustainably harvesting and producing rough-cut lumber (S_1, S_2, S_4, O_2, O_4) 2. Processing, packaging, and selling organic foods to retail food stores (S_1, S_2, S_5, O_1, O_3) 3. Seeking university food service markets (S_1, S_5, O_1, O_3) 4. Recruiting organic, grass-fed meat producers (S_1, O_1)	**W/O STRATEGIES** 1. Marketing organic foods to retail food stores (W_1, W_3, O_1, O_3) 2. Seeking value-added wood products markets for rough-cut lumber (W_1, W_3, O_2, O_4) 3. Installing more durable, more efficient reflective material on solar kiln (W_5, O_6) 4. Using a scrap wood burner as a backup energy source for the kiln (W_5, O_6)
THREATS (T) 1. New national organic food standards 2. Lack of national sustainable wood products standards 3. Competition from large companies 4. Uncertainty of outside funding 5. Insufficient number of organic growers 6. Insufficient acreage of adequate forest land	**S/T STRATEGIES** 1. Recruiting, educating, and assisting growers (S_2, T_1, T_5) 2. Recruiting, educating, and assisting land owners (S_2, T_2, T_6) 3. Seeking markets for organically grown but uncertified foods (S_1, T_1, T_3) 4. Replacing internal sustainable wood certification standards with Forest Stewardship Council standards (S_2, T_2)	**W/T STRATEGIES** 1. Seeking grants and other outside funding sources (W_3, W_4, T_3, T_4) 2. Recruiting two new marketing staff persons (W_1, W_2, T_3, T_4) 3. Beginning a strategic planning process (W_2, T_1, T_2, T_3, T_4, T_5, T_6)

Table 7.2 **TOWS Matrix for Ford**

	STRENGTHS (S)	**WEAKNESSES (W)**
	1. Committed leadership concerned for the environment 2. 100 years of providing the world with mobility solutions 3. Open, transparent communication with external stakeholders 4. Strong product design and styling capability 5. Commitment to corporate citizenship in the context of a strong and profitable business	1. Few new designs in the pipeline 2. Costs too high and declining sales 3. Inflexible factories—mass production mentality 4. Overcapacity, especially in North America 5. Business viability relies on vehicles and the fossil fuel infrastructure 6. Vehicle CO_2 emissions 7. Inadequate internal environmental and social performance measures
OPPORTUNITIES (O) 1. Mobility solutions are essential to modern civilization 2. Vehicle production is a growth business in developing nations 3. Growing demand for environmentally friendly vehicles 4. Actions to preserve the environment can reduce costs and attract investment 5. Customers want variety; "lifestyle" vehicles	**S/O STRATEGIES** 1. Assess the environmental, social, and economic impact of new factories in developing markets with stakeholders (S_3,S_5,O_2,O_4) 2. Participate in the WBCSD Sustainable Mobility Project (S_1,S_2,S_3,O_1,O_2) 3. Orient design strength to style variety with platform sharing (S_4,O_5) 4. Develop new environmentally friendly vehicles (S_2,S_4,O_3,O_5)	**W/O STRATEGIES** 1. Reduce risks and costs by managing the environmental performance of manufacturing (e.g., ISO 14001) (W_2,W_7,O_4) 2. Shift capacity from North America to China and Asia (W_4,O_2) 3. Extend ISO 14001 risk reduction and cost benefits into the supply base (W_2,O_5) 4. Delink mobility and greenhouse gas emissions (W_5,W_6,O_1,O_2,O_4,O_5)
THREATS (T) 1. NGO influence, especially regarding human rights and the environment 2. Growing demands for mobility will outstrip today's means of transportation 3. Climate change impact of vehicles 4. Regulatory environment 5. Competitors have flexible factories and product variety today 6. Intense competition and declining economy	**S/T STRATEGIES** 1. Design for the environment—use of renewable and recycled materials (S_4,S_5,T_1,T_4) 2. Initiate formal, public dialog with NGOs, jointly explore issues and solutions (S_1,S_3,S_5,T_1,T_3) 3. Exceed government schedules for fuel economy and emissions reductions (S_1,S_5,T_3,T_4) 4. Work with NGOs and governments to define and develop future markets and products (S_1,S_2,T_1,T_6)	**W/T STRATEGIES** 1. Hybrid and fuel-cell power technologies (W_5,W_6, T_1,T_2,T_3,T_4,T_6) 2. Publish annual corporate citizenship report (W_7,T_1,T_3,T_4,T_6) 3. Retool North American factories for flexible, multiplatform capability (W_2,W_3,W_4,T_5,T_6) 4. Cost reduce the design, engineering, and supply phases (W_2,W_3,W_4,T_5,T_6)

in human behavior. They say that understanding human behavior ultimately means understanding human cognitive processes, and at the heart of these processes are perceptions.[2]

In understanding the role that perceptions play in human behavior, it is important to understand that people do not operate directly on objective reality. Instead, they operate on their perceptions of objective reality. That is, people scan their environment and create a mental picture of that environment. This mental picture involves applying values, assumptions, opinions, attitudes, beliefs, and knowledge to what is observed and developing perceptions of the situation from this process. Once perceptions are formed, people make their behavioral decisions based on these perceptions. Thus, human beings employ cognitive processes (mental processes involved in having and arranging thoughts) in order to understand and respond to their environment.

Perceptions are formed via a process called categorization. Categorization is essentially a process of reducing the information from the environment and storing it in mental categories. Categories are like mental compartments used to store a range of information that is related in some way. People categorize information from their environment in two ways. Sometimes categorization is automatic; it occurs instantaneously for very familiar, overlearned information. Obvious characteristics such as color, sex, and dress are examples of the types of signals that people are likely to categorize automatically. However, people use a more thoughtful, controlled categorization process when dealing with problematic, novel, and unexpected information. Controlled categorization is complex; it requires that individuals pay close attention to information, consciously search their memory for the appropriate categories they use to interpret the information, and make conscious decisions about the information. In fact, actual perception involves both automatic and controlled categorization processes in varying degrees depending on the situation.

Furthermore, developing perceptions involves both rational conscious processes and often nonrational unconscious processes. Cognitive processing likely begins at the unconscious level, and most of what is processed probably remains at that level. Trying to consciously process all the information received from one's environment would be overwhelming and incoherent. Rather, much of the information stored in mental categories is processed at the unconscious level, with only a limited amount of that information rising to the level of consciousness. Thus, conscious perceptions result from a multitude of unconscious processes that involve both competition and cooperation among the many categories of information available.

If people's experiences were always the same, then individuals would likely develop neat categories that could be called upon individually when needed. However, human experiences normally differ from previous ones, requiring people to integrate their mental categories in a wide variety of often unique and complex ways in order to develop meaningful perceptions of their various situations. The results of these integration processes are *cognitive maps*, defined as networks of interconnected mental categories that interact in order to provide meaning and direction for the individual. Cognitive maps are used to translate experiences into knowledge and action just as road maps are used to explore the alternative routes people may want to take on a trip.

Of course, people are the basic elements that form the nucleus of all organizations. Thus, insofar as human behavior is guided by perceptions, organizations can exist only when their members share certain perceptions. They must have common perceptions of the organization's goals, of the methods for goal accomplishment, and of the roles of the organizational members with regard to goal accomplishment. In other words, organizations can exist only when the members share a common cognitive framework. From this perspective, organizations are not so much concrete entities as they are mental networks composed of shared meanings and social interactions based on common visions, goals, language, values, and so forth. The popular term used when referring to organizations as cognitive networks is "organizational culture." Shared motives, shared experiences, shared visions, shared language, shared myths, shared values, and shared assumptions all portray the cultural characteristics of organizations. With their emphasis on the word *shared*, these terms demonstrate the true cognitive nature of organizations.

Strategic Decisions: Complex, Assumption-Based, Value-Laden Choices

One of the most common behaviors in which human beings engage is decision making. People are faced with making choices all the time. Some choices are based on data that people have directly available; these choices require little more than an instantaneous, automatic categorization. Other decisions are quite complex. All the data needed are not directly available, meaning that individuals have to engage in controlled categorization processes, searching their memories for knowledge and metaphors and consciously processing the information available in making their choices.

Managers in business organizations are paid to make decisions, and the majority of the decisions that managers make are not automatic. Managerial decisions typically involve significant amounts of uncertainty and require extensive use of conscious, controlled cognitive processes. For example, decisions about the performance of employees (a standard managerial task) are normally made based on limited opportunities to observe the employees at work and require managers to search their memories for information that may be up to a year old.[3] Moving up the organizational hierarchy means facing even more problematic, uncertain issues that require managers to make decisions with increasingly significant consequences.

It is well established that values play a major role in complex managerial decisions.[4] Values are enduring, emotionally charged abstractions (categories) about matters that are important to individuals. Understanding values is no simple matter. Some have conceptualized values as existing in hierarchies; that is, some values are always more important to individuals than others. Others say that the importance of any individual value varies. They say that the real key to understanding how important values are in influencing decisions is to discern how important an individual's total system of values is to her or him. This approach views values from a holistic perspective in which the values people favor are influenced by the situation. When people actually apply their values to their decisions, both of these frameworks are likely to come into play to one degree or another. No doubt, some values are more important to people than other values; at the same time, the total strength of people's value systems and the situations in which they find themselves are also important.

Social values seem to have a particularly strong influence on the complex choices that individuals make. Social values represent a broad array of behavioral norms that are best defined by their "oughtness"—that is, social values deal with the ways people believe they should behave. Although social values may originally descend from pain and pleasure experiences, they become the social canons of a group, company, community, or society. As such, they serve to control and protect the behavior of individual citizens. This systemic support for social values gives them an existence well beyond any individual rewards or punishments. Because social values are expected within the larger group, all members usually endorse behaving in accordance with these values. This is a very powerful motivational force.[5]

Values play major roles in the cognitive processes people go through when they make complex decisions.[6] People search the characteristics of their situation, determine the consequences of these characteristics, and determine the desirability or undesirability of these consequences. Cognitive choice processes involve choosing (consciously and unconsciously) what information to pay attention to, encoding, storing, and retrieving this information, and integrating the retrieved information into a final choice or decision. Values have an especially strong influence on this process at two points. First, values are critical in helping people determine which elements of their environment are important to heed in making decisions. Second, values are the primary criteria used when individuals choose among the available alternatives. Values will become more prominent in the final choice as the decision becomes more complex, ambiguous, and judgmental.

Strategic decisions are at the pinnacle of complex choices faced by business managers. They are made at the upper echelons of the organizational ladder, and their consequences are potentially vital to organizational success. Furthermore, successfully implementing strategic decisions rests on effectively integrating them into the culture of the organization. Typically, strategic decisions must be made about issues that are ill defined, nonroutine, complex, and uncertain, and have the potential for multiple interpretations. Thus, strategic decisions tend to be unique from situation to situation and from organization to organization. They require that managers have diverse capabilities, and they require managers to integrate a variety of elements including multiple goals, multiple stakeholders, multiple decision makers, multiple attitudes, and vague time horizons. Thus, making strategic decisions in business organizations requires controlled cognitive processes. Strategic decisions are made by invoking a variety of cognitive simplification processes, anchoring final decisions on initial value judgments, and using analogies and images to identify problems and solutions. Also, understanding the strategic decision-making process requires understanding the assumptions that underlie the firm's value system. Understanding organizational assumptions means delving into the cognitive maps of the executives who make the decisions because these are the maps that represent the mental structures through which strategic choices are filtered.[7]

Sustainable Strategic Decisions: A Matter of Changing Perceptions

Thus, the models that managers use to make complex strategic decisions are the result of cognitive processes whereby the managers apply their knowledge, values, and assumptions in order to develop their perceptions of reality and to determine the directions they will take in dealing with their environment. So, why do today's business models inadequately value the greater social system and ecosystem? At the heart of the matter may be the insufficiency of current cognitive processes to account for the long-term, seemingly gradual changes that characterize most social and ecological issues.[8]

Most of the population growth and most of the manmade changes in the natural environment have occurred during the Industrial Age, but the biological evolution of the human being was essentially complete thousands of years earlier. The mental pictures that our ancestors used to comprehend their environment were developed in a very different type of world than the one that exists today. People lived in small groups in limited, stable, harsh environments. Most of their responses were geared to dramatic, short-term environmental changes. Thus, human beings developed cognitive processes that focused on the short term. Long-term thinking for our ancestors was season to season; survival was based on their responses to events that occurred daily or moment to moment. They probably perceived gradual global patterns, but they had to suppress these perceptions in order to focus on their immediate situations. For most of human history, these short-term mental processes were adequate for survival. The planet was not overpopulated, and there was safety in numbers. People had neither the potential to create long-term global changes nor the ability to deal with these changes when they occurred.

Whereas most species evolve only biologically, human beings also evolve culturally and technologically. Even though human biological evolution has occurred very slowly, cultural and technological rates of development have been nothing short of phenomenal, especially during the Industrial Age. Humans now have an exponentially growing population, a global culture based on a shared belief in the virtues of economic growth, and the technology to create long-term global changes. These are, of course, the three key factors determining humankind's impact on the planet (see Chapter 4). Unfortunately, human perceptual processes are still tied to the old world of short-term dramatic change. People still suppress long-term perceptual processes to a great extent. Thus, it is hard for humans to perceive the impacts of the global changes that are occurring. Humankind continues to reproduce rapidly even though it knows that overpopulation is one of the most serious long-term problems. Humans continue to use resources and pollute at incredible rates even though they know that there will be dire long-term consequences.

People need to take advantage of the flexibility and trainability of the human mind to achieve the changes necessary in their mental pictures of reality. Human beings are the most adaptable of the species, and they have the potential to synthesize large amounts of information. Education about the problems faced by humankind is important, but education about the way people think may be even more important. If people can understand how they learn, if they can understand how their values and perceptions influence their view of the world and their reactions to it, then they may be better equipped to modify their cognitive structures to fit the demands of their current environment.

Assumptions and Values for Sustainable Strategic Management

To summarize, strategic decisions are complex, ambiguous, and judgmental, made at the upper echelons of the organizational ladder, and they have consequences that are potentially vital to the organization's success. They are made about issues that are ill defined, nonroutine, and uncertain with the potential for multiple interpretations. As such, strategic decisions are both guided and limited by the underlying assumptions and values prominent in the organization.

As we apply these concepts to sustainable strategic management, an important question emerges. What are the appropriate assumptions and values for the effective formulation and implementation of sustainable strategic management in organizations? Of course, there is no proven single set of assumptions and values that would apply to all organizations all of the time. On the contrary, we believe that sustainability has myriad paths supported by myriad assumptions and values. All that is required is for those assumptions and values to genuinely and effectively support organizations' efforts to achieve sustainable strategic management. With this admonition in mind, we present some economic assumptions in this section that we believe account more effectively for the greater social system and ecosystem in which the economic system is nested, and we present a sustainability-centered value system that we believe supports these assumptions.

Sustainability-Based Economic Assumptions

At the heart of the assumptions held by most strategic managers are those from classical and neoclassical economics. These assumptions are heavily embedded in business school curricula, and they are reflected daily in the marketplace. Many scholars have called for fundamental changes in these assumptions, arguing that they are ultimately unsustainable and fly in the face of the entropy law. In this section, we suggest some new economic assumptions to replace the unsustainable ones.

Limiting Unlimited Growth

The most often criticized assumption of economics is that unlimited economic growth is both possible and desirable forever. Recall that we discussed in Chapter 2 how this assumption violates the basic tenets of the entropy law. In fact, many scholars have developed convincing arguments contending that unlimited economic growth is a fairy tale with the potential for a nightmare ecological ending.[9] E.F. Schumacher said thirty years ago that the first assumption of economics should be that there is such a thing as enough, and Herman Daly has consistently proposed over the years that *enoughness*—a sufficient level of economic consumption beyond which human welfare and ecological balance are significantly eroded—needs to be the primary assumption of economic theory.

Accounting for Nature

We have already extolied the problems related to humankind's over-expenditure of its natural capital. At the heart of this problem is the assumption from classical and neoclassical economics that natural capital is virtually unlimited and serves as a near-perfect substitute for manmade capital. Unfortunately, this simply does not always hold true. The money earned from the sale of hamburgers that are made from beef raised on land cleared in a rainforest is certainly not a perfect substitute for the species and forests destroyed forever so those hamburgers could be made and sold. What is needed is an assumption that natural capital is limited, precious, and often nonsubstitutable. Interestingly, two subfields have emerged in economics that operate under the assumption that nature has value and should be accounted for in economic activity. One subfield is referred to as *environmental economics*, and its basic premise is to internalize natural capital costs via regulations, taxes, and market incentives. Such mechanisms are used to assign value to resources, pollution, and wastes. However, Herman Daly and others propose a more sweeping *ecological economics*. Daly proposes a steady-state economy based on two physical magnitudes—the physical stock of capital and the flow of throughput. By adding throughput, a steady-state economy is able to operate on the reality that the entropy law imposes absolute limits on the capacity for and rate of economic activity.

Replacing Selfish Individuals with Moral People Living in Communities

Another often criticized economic assumption concerns the relationship between self-interest and the common good. In the economic theory of Adam Smith, self-interest is the infamous *invisible hand* that efficiently guides the market system toward a fair allocation of resources. Although Adam Smith himself made a clear distinction between self-interest and selfishness, economic self-interest today is most often interpreted as the selfish pursuit of individual goals. This assumption of selfish individualism is supported because, in theory, maximizing individual wealth is the most direct route to maximizing the welfare of society. Another key dimension of this assumption is that individuals are essentially value neutral except for those values that can be expressed monetarily in the market. New economic models counter this assumption, saying that while humans certainly make rational individual economic choices designed to satisfy their needs and desires, they make these choices within the moral dimensions of a meaningful community structure.

Values for Sustainable Strategic Management

Of course, as we have pointed out repeatedly, the integration of these assumptions into a firm's strategic management processes will require an organizational value system that supports them. We discussed in Chapter 3 that sustainability is an ideal core value to support these assumptions because it can transcend the divergent dilemma facing humankind today—balancing economic activity with social system and ecosystem viability. Recall from our discussion that a core value is not viable if it exists alone. Rather, it must be supported by a set of instrumental values that facilitate its implementation. In this section, we offer eight instrumental values that we believe allow for human decisions based on sustainability—wholeness, posterity, community, appropriate scale, diversity, quality, dialogue, and spiritual fulfillment.[10]

Wholeness

Conceptually, wholeness is about interconnectedness, relatedness, balance, mutual causality, and the connections between past, present, and future. Wholeness reminds humans that they are products of their collective past and architects of their collective future. It reminds them to connect with all elements of their environment and to consider their impact on these elements in the decisions they make. Thus, a value for wholeness helps strategic managers to recognize that all of the earth's living subsystems are parts of a supranational ecological system. Wholeness also helps strategic managers to remember that survival depends on successfully interacting with the other living subsystems on the planet because the whole cannot survive if its parts are destroyed. Furthermore, insofar as the whole is defined by how its parts interact with one another, valuing wholeness will help strategic managers to better perceive and attend to the relationships with other elements of the environment.

There are two reasons that valuing wholeness is advantageous for a sustainable balance among economic success, social responsibility, and ecosystem viability. First, a value for wholeness can provide business organizations with the broad perspective they need to remain competitive in today's global network of business activity. Thinking of the world as an interconnected whole can provide a much clearer understanding of the coevolutionary relationships that organizations need to develop with their stakeholders, and holistic thinking is often essential to the novel, creative efforts necessary to develop new products, new services, and new processes. Second, valuing wholeness encourages strategic managers to consider the impacts of their decisions on the greater social system and ecosystem by allowing them to understand how the organization fits into, affects, and is affected by those systems.

Posterity

"We didn't inherit the earth from our parents. We borrowed it from our children." This well-known ancient proverb clearly describes why posterity is an important value for achieving sustainability. Valuing posterity, believing that future generations of human beings and other species are prominent factors to be considered in humankind's decisions, is instrumental in attaining a sustainable economic, social, and ecological balance.[11] A value for posterity can be an important ingredient in effectively managing the change and turbulence that all organizations now face and will continue to face. Adopting posterity as a value encourages business organizations to develop a vision of what they are and what they want to become. Clear visions of the future are critical for organizational success. Visions serve as common denominators around which strategic decisions are shaped and implemented. Shared visions in organizations encourage employees to think strategically, and when strategic thinking is a part of an organizational culture, the company is better prepared to manage its opportunities and threats in ways that are advantageous to its survival and prosperity. In addition to supporting economic success, posterity is also important for achieving ecological and social sensitivity in organizations. Taking future generations into account in strategic decisions will significantly influence a wide range of choices. If strategic managers believe that clean water, clean air, abundant resources, biodiversity, natural beauty, economic justice, human rights, and employee rights are the birthrights of all generations, then the decisions they make are bound to better reflect a concern for the earth and its people. The Iroquois Nation had a seven-generation planning horizon; the nation's leaders tried to predict the effects of their decisions for the next seven generations to follow. This type of planning horizon for business organizations would tremendously enhance the prospects of achieving sustainability.

Community

As we discussed above, many social and ecological problems stem from the neoclassical assumption of radical individualism, the belief that selfishly serving one's own interests will necessarily result in the collective good. We are fostering the opposite notion here: individuals, organizations, and economies are parts of a greater community, and thinking selfishly can lead to individual actions that are detrimental to the common good. Communities are cognitive networks of individuals, organizations, and institutions that often share a common geography and always share common values and aspirations. From these common values and aspirations come the cultural mores and ethical systems that guide the actions of community members, including the actions taken by business organizations.

Strategic managers who value the greater community are better equipped to make decisions compatible with achieving sustainability. Valuing the more comprehensive communities to which they belong helps them to be more aware of the interconnections between their decisions and the quality of life in the communities where they operate. From this, strategic managers are better able to recognize that their organizations can prosper over the long run only if the community can maintain a balance between a healthy natural environment, ample opportunities for human development and fulfillment, and a healthy system of economic activity. Accordingly, strategic managers who value community will likely benefit from numerous economic advantages, such as customer loyalty, a positive public image, and employee commitment, while contributing to the protection of society and the natural environment. They are also more likely to become involved in the community-based decision-making structures that many scholars believe are critical for protecting the commons.[12]

Appropriate Scale

Sustainability is a matter of scale.[13] As we have discussed previously, humans live on a small planet, one that is becoming overburdened with population and economic activity. A value for appropriate scale concentrates organizational efforts on issues related to this situation, such as resource reduction, materials reduction, energy efficiency, recyclability, and reusability. Thinking in terms of appropriate scale is no easy task for people in today's world because the persistent pursuit of unlimited economic growth leads to an attitude that bigger is always better. Nonetheless, appropriate scale is clearly a value with a huge upside regarding its potential for providing both economic and ecological benefits to both organizations and society.

Basic to appropriate scale is the amount of energy and resources transformed from their natural state into outputs, including wastes. Thus, a value for appropriate scale has implications for every aspect of the economic cycle. At the production end of the cycle, appropriate scale will help managers more accurately account for the scarce natural capital that forms the foundation of all economic capital. A value for appropriate scale will encourage strategic decision makers to implement policies aimed at using as little as possible of the earth's nonrenewable resources. Organizations applying appropriate scale are more likely to focus attention on searching for ways to save energy and to use more renewable energy sources in the production and delivery of their products and services. Appropriate scale will also encourage strategic managers to look for ways to reduce the materials that go into their products, including packaging, and to search for safe substitutes for toxic and dangerous materials in their products, services, and processes. Valuing appropriate scale in the production of goods and services also has tremendous implications for the technologies used by organizations. Production technologies based on more efficient type II and type III industrial-ecosystem models (discussed in Chapter 2 and Chapter 5) would emanate more readily from such a value. Creating renewable energy technologies, resource conservation and recycling technologies, inter-organizational waste-resource loops, and so on would become serious agendas for organizations. Appropriate scale has a human dimension as well. To be appropriate, the scale of operations should allow for jobs structured to provide good work, fairness, and loyalty for employees.

Diversity

Research has shown that diversity is critical for maintaining the ecosystems that support life, for providing the necessary linkages for social and cultural survival, and for maintaining successful organizations in today's global economy.[14] James Lovelock's Gaia hypothesis sprang from his discoveries that as biodiversity increases, the planet's conditions for supporting life become more favorable. The literature supporting the relationship between cultural diversity and social vitality is overwhelming. Of course, in today's global economy, diversity is especially critical to the success of business organizations. Not only can encouraging and effectively managing diversity help organizations avoid negative outcomes like increased turnover and legal problems, it can also help them improve their talent pools, and it can contribute significantly to improved competitiveness via better market understanding, higher-quality problem solving, improved leadership, and more effective global relationships. Thus, diversity contributes ecologically, socially, and economically to sustainability.

Quality

Robert Pirsig engaged in a fascinating search for the true meaning of quality in *Zen and the Art of Motorcycle Maintenance*. His schizophrenic inquiry led him to discover that quality does not result from something people do or see.

It is not an objective measure. Rather, it is a perception with deep cognitive roots. This conclusion disturbed him; quality is, no doubt, a subjective mental image, but its basic purpose is to improve objective awareness of the world, that is, to provide observable criteria for making choices. How can it possibly be both subjective and objective? His final breakthrough (as well as his mental breakdown) came when he realized that quality could not exist either solely as a subjective mental picture or solely in the objects of the real world; quality was neither object nor subject but a unique event that existed in the interactions between the two, between the mind and its surrounding environment.

This is the same perspective on quality that has permeated business organizations since the quality revolution of the 1980s. Quality is no longer the exclusive bailiwick of quality control engineers applying absolute objective standards, and quality is not just a subjective ideal, often hatched in the minds of advertising executives. Instead, quality is determined through interactions between managers, operational employees, customers, suppliers, quality control personnel, and others. Quality according to this definition is an overall perception of what the firm's products and services should be. As such, quality serves as the guiding force behind the firm's operations and its relationships with its stakeholders.

As an instrumental value that supports sustainability, quality integrates many of the elements of wholeness, posterity, appropriate scale, and community. That is, valuing quality means valuing networks over dominance, tomorrow over today, and better over more. Once organizations adopt a philosophy that how well products are made and how well customers are served are more important than how many products are produced and how many customers they are sold to, then the proper scale of their operations can be defined by something other than physical growth. When quality is the nucleus around which organizations revolve, they are likely to adopt a scale of operations that allow them to focus on developing individual relationships within their stakeholder network. Improved customer loyalty, more stable supplier relationships, more participative interactions among organizational members, and improved operational efficiency are all possible outcomes for organizations that adopt quality as a key value in their strategic decision-making processes.

A value of quality will best support sustainability if it includes three basic dimensions—quality of products and services, quality of work, and quality of life. Quality products and services improve customer loyalty, increase efficiency, and so forth, and they also support sustainability because they last longer, are worth repairing, and can be exchanged more readily in second-hand markets. Attaining the sustainability promised by focusing on quality products and services is not possible unless organizations also value quality of work over quantity of work. Quality products and services are simply not possible without quality work. Structuring jobs that satisfy human needs as well as organizational needs can no doubt improve the quality of products and services. Moreover, the psychological satisfaction that people derive from good work can serve to curb their desire to seek satisfaction via the consumption of more and more goods. Finally, valuing quality of life is important for achieving sustainability via quality. This encourages strategic managers to recognize that all of their stakeholders have rights to physical well-being, long-lasting happiness, personal fulfillment, and a hopeful future. Valuing quality of life brings a wide variety of economic, social, and environmental issues to the attention of organizations, including job design, organizational reward systems, employee health and safety, shareholder wealth, community economic development, pollution, waste control, and so forth.

Dialogue

The term "dialogue" refers to the Socratic process designed to explore issues and ideas until their underlying assumptions, values, and principles are revealed, exposed, and, if necessary, changed. Via dialogue, organizations are capable of creating interaction patterns that allow underlying assumptions and values to surface openly and be questioned. Using dialogue as the basis for interaction with both internal and external stakeholders puts organizations in positions to realistically assess their perceptions concerning their employees, their community, and the planet on which they do business. Thus, via dialogue, organizations can establish with their stakeholders the kinds of communication processes that can be very instrumental in sustaining a healthy ecosystem-social system-economic system balance. Valuing dialogue will encourage firms to establish mechanisms that allow open discussion with all stakeholder groups, and it will encourage firms to participate in established community organizations and processes designed to benefit the community and protect the planet around them.

Spiritual Fulfillment

As elusive as it seems, there is remarkable agreement about the meaning of spiritual fulfillment in most of the world's great philosophies, religions, traditions, and mythologies. To be spiritually fulfilled is to experience states of peacefulness, love, joy, happiness, enlightenment, satisfaction, accomplishment, and creative expression. Schumacher said that these are the goals of moving *higher* (see Chapter 2), and Daly said that they are *ultimate ends*—outcomes that have no instrumental value, only intrinsic and sacred value.[15]

The economic goals of organizations have typically been seen as ends in and of themselves, and the relationship between organizations and their stakeholders has primarily been an economic one. Organizations seeking to meet the higher-level spiritual needs of humankind have traditionally been relegated to positions outside the economic arena. The result is an economy made up largely of organizations designed to serve no purpose beyond providing economic prosperity and physical comfort.

Many now say that kindling the flames of the human spirit is critical for modern organizations.[16] They say that organizations need to focus more clearly on their roles in contributing to the quality of life in the larger community, and they need to create structures, processes, and outputs designed to fulfill the spiritual as well as economic needs of the humans whose lives they touch (employees, customers, etc.). Valuing spiritual fulfillment allows organizations to put economic success, social responsibility, and ecological protection in their proper perspectives as avenues toward the realization of a higher quality of life. As an instrumental value for sustainability, spiritual fulfillment can provide the mental pathways that lead individuals and organizations beyond material consumption and wealth to a higher level of satisfaction and purpose. This is essential if humankind is ever to truly accept a critical tenet of sustainability: finding joy in doing more with less.

Standing for Sustainability

As we come to the end of this section on the formulation of sustainable strategic management, we want to summarize some important points from this and previous chapters and to weave these points into a model that will provide an overall picture of the sustainable strategic management formulation process as we see it. We begin our summary with the concept of enterprise strategy discussed in Chapter 3. Recall that enterprise strategy is the level of strategy at which the value systems of managers and stakeholders come into play in concrete terms. Also recall that understanding a firm's enterprise strategy means analyzing the interactions among three components—the values held by the firm, the social and environmental issues faced by the firm, and the key stakeholders of the firm. Such an analysis helps firms to clarify what they stand for, and to understand the underlying ethical foundations of their strategic choices.

We have discussed each of these components in some depth as they relate to sustainable strategic management in this and previous chapters. In this chapter we have suggested a set of eight instrumental values to support a

Figure 7.1 **Sustainable Strategic Management Process**

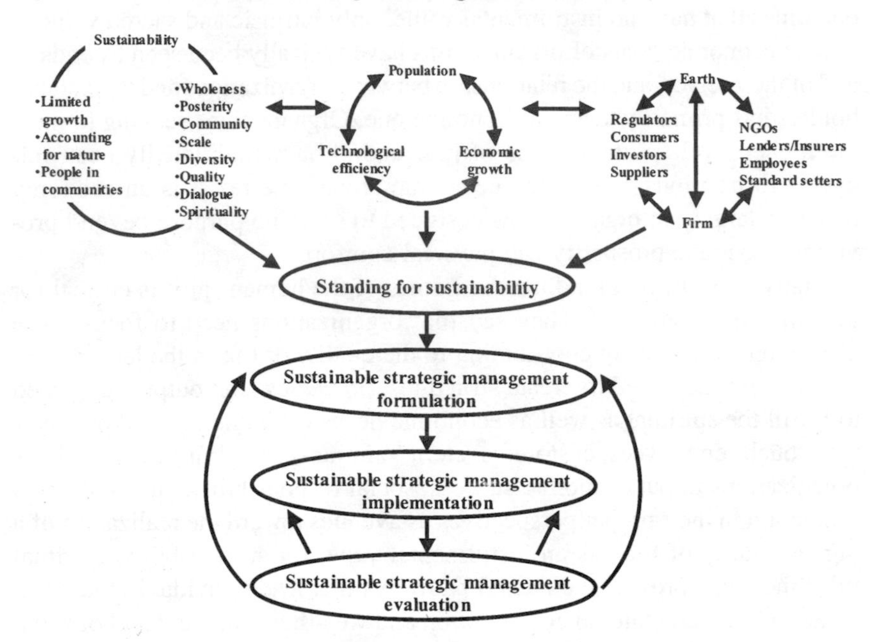

core value of sustainability—wholeness, posterity, community, appropriate scale, diversity, quality, dialogue, and spiritual fulfillment. In Chapter 4, we discussed myriad social and ecological issues facing business organizations that result from the interaction of economic activity, population growth, and technological advancement on the planet, including resource depletion, degradation, pollution, climate change, species loss, human health problems, reduced quality of life, and so forth. In Chapter 5 we discussed how many stakeholders take an active interest in the ecological and social activities of organizations today, including the government, consumers, investors, suppliers, lenders, insurers, nongovernmental organizations, employees, and industry standard setters.

Our contention here is that organizations that apply the value system outlined herein (or other sustainability-centered value systems) to the decisions they make regarding the social and ecological issues they face, the stakeholder interests they serve, and the best means to serve those stakeholder interests can be said to "stand for sustainability." As such, the streams and patterns of strategic decisions that emerge in these organizations will by definition be consistent with the tenets of sustainable strategic management. That is, these organizations are setting in motion strategic management processes that allow them to make their strategic decisions and measure their success in economic, ecological, and social terms. Figure 7.1 depicts this process.[17]

Conclusions

Relativity and quantum physics have vanquished the idea that everything from the structure of the universe to the behavior of human beings is a mechanical process governed by deterministic laws. We now know that people can view their world only through their perceptions. Perceptions result from complex mental processes that involve filtering environmental data through intricate cognitive frameworks in order to ascertain meaning and choose courses of action. These cognitive frameworks are assumption-based and value-laden, especially in the complex, uncertain, unfamiliar, ill-structured situations faced by strategic managers in business organizations.

Adequately including the greater social system and ecosystem in the mental pictures that guide strategic managers in their strategic decisions can be achieved only by refocusing their mental frameworks to include gradual, long-term processes. Doing so will require changing the way managers think.

There are many relevant assumptions and values related to sustainable strategic management. We believe that assuming there are limits to growth, that there are no perfect substitutes for social and natural capital, and that humans live in communities and have values beyond the economic realm are all important for sustainable strategic management. Further, we believe that these assumptions can be brought to life in business organizations if they are supported by the instrumental values discussed in this chapter (or some related value system). If strategic managers will apply these values to the decisions they make regarding the issues they face and the stakeholders they serve, then the strategies they formulate will be consistent with sustainable strategic management. Once these strategies are formulated, the next question is how to implement them.

Part III

Implementing and Evaluating Sustainable Strategic Management

The first seven chapters of this book have made the case for and provided an in-depth explanation of what sustainable strategic management (SSM) is and how SSM strategies are formulated. Of course, as with any strategic management effort, effective formulation must be accompanied by effective implementation and evaluation. In this part of the book, we will highlight a number of salient SSM functional level strategies, systems, and processes necessary for effective implementation and evaluation of sustainable strategic management. In Chapter 8, we will cover the functional level strategies associated with the primary activities of the closed-loop value chain, and in Chapters 9 and 10 we will discuss those functional level strategies associated with the support activities of the closed-loop value chain. As mentioned several times in Part I and Part II, SSM strategies provide the means to achieve the organization's stated sustainability goals, thus effective implementation and evaluation of these strategies are critical for fulfilling visions of a sustainable firm in a sustainable world.

In Chapter 11, we revisit Anthony Flaccavento, Bill Ford, and the organizations they lead—Appalachian Sustainable Development and Ford Motor Company, respectively. We discuss in some depth the SSM strategies these firms have followed, and we use the examples of these two firms to draw conclusions regarding where humankind is in its search for sustainable strategic management and where it needs to go in the future in order to make sustainable strategic management a global reality.

8

Systems For Sustainable Strategic Management

As we begin this chapter, it is important to point out that, while formulation, implementation, and evaluation are often seen as separate functions that occur in a sequential, orderly fashion, the reality is that sustainable strategic management (SSM) requires blending these three components on an ongoing basis. Formulation, for instance, often includes some attention to the how of implementation. For example, organizations often incorporate pollution reduction technologies into their pollution prevention plans. Similarly, assessing an organization's progress toward its sustainability goals can be a critical step in determining whether these goals need to be changed or not. For instance, if a supermarket's program of recycling cardboard boxes is working satisfactorily (implementation and evaluation), it may want to expand the program in the near future to include plastic bag and other container recycling programs (formulation).

With this admonition in mind, we will begin our discussion of SSM strategy implementation. Implementing a plan (whether for sustainability or other purposes) is usually referred to as operational planning, a process that provides the details of how SSM strategies will be carried out.[1] Specifically, operational planning entails the establishment of organizational objectives, which are short-run, time-sequenced, measurable performance targets that support long-term goals. These objectives serve as guidelines for the actions of organizational members as well as providing indexes for measuring organizational performance. In sustainable strategic management, objectives such as specific reductions in emissions or improvements in working conditions are established within the context of organizational sustainability goals (see Chapter 3). After objectives are set, action plans, policies, and procedures that support SSM strategies are established to achieve the stated objectives.

Linking the budgeting process to SSM strategies allows the allocation of resources that enables the accomplishment of SSM strategy objectives. Designing reward systems supportive of SSM strategies and installing adequate information systems, performance tracking systems, and controls are also instrumental in effective SSM strategy implementation. Furthermore, strategic managers will need to facilitate the creation of learning organizational structures and cultures supportive of the core value of sustainability.

As discussed in Chapter 5, core competencies (unique strengths) are found within the functional level of the firm. SSM functional level strategies in such areas as marketing, finance, operations, and so forth are designed to develop and exploit these core competencies all along the type III closed-loop value chain. These core competencies form the cornerstones of SSM competitive level strategies. A sustained competitive advantage is the result of exploiting the firm's core competencies in the marketplace. Thus, effective SSM strategy implementation requires instituting SSM functional level strategies that develop and exploit the core competencies on which the firm's SSM competitive level strategies are based. This and the next two chapters will cover the functional level systems and strategies that are essential in the effective implementation and evaluation of SSM strategies. In this chapter we will begin with a discussion of sustainable management systems, and then we will discuss those functional level strategies and systems associated with primary activities of the type III closed-loop value chain, including research and development, procurement, operations, facilities, and marketing.

Sustainable Management Systems

At the heart of implementing sustainable strategic management in organizations is the development of integrated sustainable management systems (SMSes)—traditionally referred to as environmental management systems (EMSes) or environmental health and safety management systems (EHSes). The shift from EMSes to SMSes is necessary because successful sustainable strategic management requires expanding the scope of EMSes to fully account for the social dimension of sustainability, thus becoming an integrated system that focuses on all three dimensions of sustainability—economic, social, and ecological.

Depending on the nature of the firm, sustainable management systems vary both in terms of what elements are included in them and the relative importance of each element. However, there is general agreement on some of the important elements necessary for an effective SMS: At the heart of an effective SMS is strong commitment to improved sustainability performance at the board and senior management levels of the organization. An effective SMS also requires developing and communicating throughout the organization clear sustainability goals along with specific objectives and targets to support these goals. Critical to any SMS is the development of core competencies necessary to implement SSM strategies designed to achieve the outcomes outlined in the firm's sustainable strategic management goals and objectives. Systematic reviews of sustainability performance measured against these stated goals and objectives as well as against the firm's past sustainability performance and regulatory requirements are also necessary elements of an SMS. Further, these reviews can be carried out only if effective sustainable information and reporting systems are established. Of course, none of this will take place unless human resource processes, including training, employee performance standards, performance appraisal systems, and reward systems, are brought into line with the firm's sustainability performance goals. All of this requires participative, empowerment-based structures that encourage all employees to take direct responsibility for sustainability improvements as basic parts of their jobs.

Few organizations to this point have included the firm's social performance in their EMS, thus creating an integrated SMS. However, it is critical to implementing SSM strategies that all three sustainable performance measures be included in a comprehensive SMS. As in strategy formulation, the key to implementing an effective SMS is to utilize the collective wisdom of the firm in a generative formulation process. This will allow the SMS to reflect the shared vision of sustainability within the organization. The type III closed-loop value chain portrayed in Figure 5.3 provides an excellent conceptual framework for establishing a comprehensive SMS.

A useful model to use in designing an effective SMS is the Natural Step because it promotes a need for a common understanding of ecological connections.[2] The Natural Step is an international educational organization whose mission is to accelerate society's movement toward sustainability. The model is built upon the fundamental principles of nature, applying them to business decisions. The principles of the Natural Step specify system conditions that represent the components of sustainability. The four conditions are as follows: (1) nature's functions and diversity must not be subject to increasing concentrations of substances extracted from the earth's crust; (2) nature's functions and diversity must not be subject to increasing concentrations of substances produced by society; (3) nature's functions and diversity must not be impoverished by overharvesting or other forms of ecosystem manipulation; and (4) resources are to be used fairly and efficiently to meet basic human needs worldwide. EMSs address the issues of environmental management as set out in all four conditions, but they do not address the fairness aspect of meeting basic human needs set out in the fourth condition. However, SMSs based on frameworks such as the Natural Step focus on both ecoeffectiveness and socioeffectiveness, addressing the gaps between the developed and developing worlds and making positive contributions to both natural and social capital.

Sustainable Research and Development Systems

If much of human existence and progress can be said to be the result of ideas, then organizational research and development may be considered the brains of SSM strategy implementation. Here products and services are initiated, nurtured, manipulated, and tested. In addition to overall effectiveness, cost, and other typical product or service characteristics related to research and development activities, sustainable strategic management also focuses on traits such as dematerialization, resource intensity, recycled content, nontoxic components, ease and safety of use, durability, aesthetics, social impacts, and so forth. The key to research and development in sustainable strategic management is to use nature as a model for the research and development system. Systems that mimic nature are effective systems that consider waste as lost profit and create value by design. Nature uses feedback to close the loop and trigger adaptations that lessen physical constraints. Research and development systems structured around a model of nature provide the framework for systems thinking and an opportunity for organizations to view limits as opportunities for new product development, innovation, and entrepreneurial learning.[3] Three particular research and development tools are highlighted in this chapter that can aid strategic managers in developing systems that mimic nature. Two were briefly introduced in Chapter 6—life-cycle analysis (LCA) and design for sustainability (DFS); the third is information substitution for energy and materials.

Life-Cycle Analysis

Remember that LCA is a total systems approach designed to assess the environmental impacts of an organization's product or service at multiple stages of its existence.[4] LCA involves analyzing resources, emissions, energy, and environmental effects all along the closed-loop value chain. With LCA, not only do organizations have a tool that can give them solid data on how to improve their environmental performance and reduce resource intensity, they also have a tool that can help them to extend the life of their products, making them more competitive in the marketplace.[5] Recall from Chapter 6, SSM strategies can enhance profitability by reducing embodied energy, increasing durability, and lowering resource intensity of products and services. LCA provides an excellent tool to assess these factors and reduce their environmental impacts.

LCA begins in nature. That is, in the first stage of LCA, an analysis of the type and amount of raw materials and energy used and the ecological impacts of acquiring the raw materials and energy that serve as inputs into the firm's products and processes is performed. In the second stage, the ecological impacts of the manufacturing process are assessed. This includes examining the materials and energy used in component manufacturing processes, final product manufacturing processes, and product assembly processes. Third, the transportation and distribution systems related to delivering the product to market are analyzed in terms of distribution modes, distances, fuel consumption, and so forth. Fourth, the environmental consequences of how the product is used are analyzed, including assessments of product durability, energy requirements, polluting potential, and the like. Fifth, the product's potential for reuse or recyclability is analyzed. The sixth stage of LCA is to examine the product's ultimate disposal in terms of its toxicity, volume, biodegradability, and so forth.[6] The result of these analyses can be grouped together, for instance, by media (e.g., land, air, and water) or aggregated to form a total score, and then compared with past efforts, with peer products or services, or with some external standard or benchmark.

Business organizations, governmental agencies, and certification organizations all over the world are making concerted efforts to base their environmental actions on effective LCA. The result has been a mushrooming of LCA tools and techniques, such as LCA databases, expert systems, and software. Some businesses, including auto manufacturers Volvo in Sweden and Fiat in Italy as well as chemical manufacturer Henschel in Germany, have reportedly fully integrated LCA into their respective processes. These firms use LCA as a way of thinking strategically and as a means of learning about how their organizations are affecting the natural environment. Many other organizations have conducted extensive LCA as a means to achieve an entire series of voluntary standards of the International Organization for Standardization (ISO), leading to ISO 14040 certification; such firms include the Canfor Company (forest products), Hewlett-Packard, Scott Paper, and Procter & Gamble. Besides ISO certification, other reasons that LCA may continue to gain in popularity in the future are its potential uses in product take-back programs and in voluntary environmental agreements.

It is important to understand that LCA is an extremely complex concept that is difficult to implement. LCA outcomes are easily skewed by the assumptions of those doing the analysis, and LCA seldom provides clear-cut answers concerning the environmental benefits and challenges of products and processes. Although LCA technology is improving rapidly, it is still difficult to conduct LCA on products that are very complex in terms of materials, components, and design. Other problems with using LCA are that the process can be very information intensive and, therefore, cost intensive, depending on how many factors are assessed and the extent of upstream or downstream activities within its scope. After the data are gathered, assessment issues include the need to compare differential impacts (e.g., energy demand with water pollution), the need to keep the information confidential, and the need to determine the potential for recycling.[7]

Regardless of the difficulties associated with LCA methodology, forging ahead with its development and improvement is absolutely essential. LCA is indispensable if the laws of thermodynamics are to be effectively included in product and process design and development. Even though the methodological limitations of LCA mean that organizations cannot currently get all the answers they need about the impacts of their products and services on society and the planet, at least LCA is philosophically consistent with searching for and eventually finding those answers. Fully implementing sustainable strategic management will require the expansion of LCA to account for the social as well as the ecological impacts of the firm's products and services all along the type III closed-loop value chain. Thus, factors such as the impacts on communities of extracting resources or the impacts of the firm's manufacturing process on its human capital should be included in LCA in sustainable strategic management.

Design for Sustainability

Design for Sustainability (DFS) is another research and development tool that enables strategic managers to think beyond "what is" to "what can be" in terms of designing innovative products and services for both the developed and developing markets of the world. At present, because social criteria have not been effectively integrated into DFS, most firms are utilizing Design for Environment (DFE), focusing primarily on ecological considerations. The idea behind DFE is to ensure that all relevant and ascertainable environmental considerations and constraints are integrated into a firm's product design processes. The goal of DFE is to implement environmentally sound production processes and to produce environmentally sound products while remaining competitive in terms of product performance and price. There is a wide range of dimensions that encompass DFE, including: design for disassembly, refurbishment, component recyclability, materials recyclability, material substitution, source reduction, waste reduction, product life extension, remanufacturing, and energy and materials recovery.[8]

Implementing DFE requires that organizations engage in two broad categories of activities. The first category involves an array of comprehensive and cross-functional endeavors. These include: reviewing internal documents to determine what unnecessary environmental damage is being caused by the firm's product components and processes; reviewing specification requests from customers to ascertain whether or not they may be ecologically unsound; and identifying unnecessary steps in the production process. Also included in this category of broad, cross-functional DFE activities is the analysis of upstream and downstream wastes. Analyzing wastes from their source allows organizations to "learn from waste."[9] The second category of DFE activities is comprised of those aimed at analyzing specific design options, products, processes, or inputs. These analyses begin with LCA, and they generally require three analytical stages. In the first stage, called scoping, the targeted design option, product, process, or input is identified, and the depth of the analysis necessary to complete the DFE is determined. In the second stage, data gathering takes place. In the third stage, the data are carefully translated so that they can be used by design teams.[10]

DFE encourages managers to embrace the view that all products and services are part of nature. Applying this idea to the design of products and services means recognizing that two discrete metabolisms exist on the earth: the biophysical cycles of nature, and the technical cycles of industry. As discussed in Chapter 5, if products are designed and manufactured based on biological and technical nutrients, they would either biodegrade, providing nutrients for biological cycles, or remain in the closed loop, where they perpetually circulate as valuable nutrients for industry. Designing products that isolate technical and biological nutrients allows them to retain their quality in the closed-loop value chain rather than being recycled into lower quality products or discarded as unusable wastes. Thus, with the right design, all products and materials manufactured by industry can safely feed the planet's biological and technical metabolisms, providing nourishment for something new.[11]

Hewlett-Packard (HP) is an organization that claims leadership in the DFE arena. HP's DFE guidelines address both generalized criteria, such as reducing the number and types of materials used and increasing the use of recycled materials, and specialized criteria, such as eliminating the use of toxic flame retardants and the use of molded-in colors and finishes instead of paints and coatings. In doing so, HP attempts to follow the ISO plastics labeling standard, to qualify more than 400 of its office products under the U.S. Environmental Protection Agency's Energy Star energy efficiency program, and to meet design, energy, noise, and ease of disassembly criteria for German Blue Angel certification (discussed briefly in Chapter 6). Another prominent DFE example is Apple Computer's design process for the Power Mac G4 desktop computer. Design decisions were made that minimized the computer's power needs, lowered its thermal profile, eliminated components, eliminated its need for flame retardant chemicals, improved its expandability and upgradability, and improved its ease of disassembly. Mitsubishi Electric Corporation's Nagoya Works plant involves all 5,000 employees and independent contractors in its DFE efforts. The firm uses DFE in both new product development and current product improvement. Every new product or product improvement is required to be screened with regard to several criteria: quantity of resource and material inputs, use of renewable resources, energy efficiency, safety, ease of disassembly, durability, packaging, and information disclosure. Each new or improved product goes through three design reviews at both the product design stage and quality assurance stage of the production process in order to ensure that the expected design improvements are actually met.[12]

One of the more promising current trends in DFE is the emergence of *green chemistry*, which involves the search for and development of environmentally benign chemical synthesis and processing. Whereas DFE efforts traditionally focus on the mechanical design of products and processes, green chemistry provides a more fundamental focus on materials at the molecular level. At this level, there is a greater chance of identifying major product and process breakthroughs that can lead to radical fundamental solutions to DFE problems. Green chemistry has the potential to improve separation and extraction methods, provide more environmentally safe catalysis methods, increase the use of renewable resources, improve product biodegradability, improve feedstock regeneration from waste byproducts, and so forth.[13]

As mentioned earlier, fully implementing sustainable strategic management will require that DFE expand its focus to include the social dimension in the design process, thus becoming DFS. This means that sustainable strategic managers will need to expand the scope of their question, "What can be?" By including the social dimension in this question, sustainable strategic managers are not only questioning how they design their products, services, and processes; they are also asking themselves what kind of world they want and what kinds of contributions their products, services, and processes make to the social and natural capital of that world.

Substituting Information for Energy and Materials

The final sustainable research and development tool highlighted here is the substitution of information for energy and materials. Remember that physical resources decline with use and are subject to the entropy law; the more you use, the less you have. On the other hand, information resources are regenerative in nature; the more you use, the more you have. As in nature, information beats scarcity. Living systems adapt as limits are approached, and information becomes embedded in their structures, resulting in more efficiency and effectiveness. The advantage that information has over low-entropy natural resources is that the more it is used, the less likely it will be to run out.[14] The irony is that information has long been an energy and material substitute. Any kind of communication that makes available information to reduce the need for extra (wasted) energy and materials can be considered a sustainability-related information-based substitution. While computers have revolutionized the way society views and uses information, information-based substitution actually began well before the electronic revolution. For example, there is no doubt that print advertisements and telephone directories have helped save huge amounts of energy over the years by allowing customers to communicate over long distances without traveling, by helping people avoid wasting energy driving to the wrong locations, and so forth.

However, many contend that the advent of electronic information systems, including computers, telecommunication networks, and microelectromechanical devices, provides organizations with virtually unlimited opportunities to substitute information for energy and materials. These advanced information systems, through their hyper sensing and real-time transmission capabilities, can provide organizations information about their products, services, and related activities that can help to minimize the amount of energy and materials necessary for their production, consumption, and distribution. Among the numerous examples of such sustainability-related information technologies are programmable (timed) building HVAC thermostats that reduce heating and cooling use, sound-and-motion lighting detection systems that turn off room lights when rooms are empty, and computer-assisted design and manufacturing systems with virtually zero tolerances that allow for significant reductions in raw material wastes. With these technologies, information is being automatically collected, processed, and disseminated to the appropriate decision sources so that energy and materials consumption are minimized.

A good example of substituting information for energy is the approach of the Warehouse Group, which operates more than 100 retail outlets in New Zealand. Its energy management team designed an environmental management software system that automatically controls the heating, air conditioning, and lighting of the whole retail chain from its central office in Auckland. The software provides controllers with three-dimensional energy consumption patterns for each store, measures energy use every half-hour, and e-mails store managers on both periodic and episodic bases to ensure that energy equipment and energy budgets are working as planned. During the first seven years of operation, this system has allowed Warehouse to cut in half its energy consumption per square meter while significantly expanding its operations.

LCA, DFS, and information substitution are three important SSM functional level strategies that can be used in developing core competencies that support either low-cost or market-differentiation SSM competitive strategies. These tools provide the means to reduce the environmental impacts and the resource intensity of products as well as the means to design products that mimic the processes of nature. Thus, they provide avenues for fitting organizations into the greater social system and ecosystem in sustainable ways (Figures 1.2 and 5.3), and in doing so, open a whole new world of economic opportunities for organizations to pursue in their quest for sustainable strategic management.

Sustainable Procurement Systems

The sustainable procurement of supplies (which begin as low-entropy natural resources) is instrumental in implementing SSM strategies that focus on preserving the planet's natural and social capital. When creating sustainable procurement systems, it is critical to design them around the type III closed-loop value chain. Product stewardship begins at the low-entropy, resource end of the closed-loop value chain. Thus, in sustainable strategic management, the overriding goal of purchasing managers must be to minimize the ecological, economic, and social costs of their resource acquisition strategies. Simply stated, socially and environmentally sensitive products and services can be produced only with socially and environmentally sensitive inputs; thus, as discussed in Chapter 5, sustainability-based supplier relationships are critical for firms seeking to practice sustainable strategic management. Sustainable procurement systems function to provide usable tangible and intangible sustainable inputs to the organization. Raw materials, semi-finished goods, products ready for assembly or packaging, and the full range of contracted services are all potential areas covered by sustainable procurement systems. Within these systems, the content, manufacturing, and delivery of procured products and services should be assessed for desirable sustainability characteristics.[15]

Sustainable Purchasing

The first step in sustainable procurement is to integrate sustainability criteria into purchasing decisions and to determine which suppliers meet the firm's sustainability expectations.[16] As discussed in Chapter 5, at the heart of developing effective sustainability-based supplier relationships is conducting regular supplier audits of the supplier's social and environmental performance. Recall that Shell uses such audits to monitor the child labor practices of its suppliers and that the apparel and textile manufacturers in the United States have developed a set of standards for working conditions, wages, and other terms of employment for their suppliers in developing countries.

Other firms require suppliers to secure ISO 14001 certification and to adopt pollution prevention policies. General Motors and Ford were among the first ISO 14001 certified large firms to apply this sustainable supply chain approach to all of their suppliers. Others, such as the Japanese auto manufacturer Toyota, have taken a more targeted approach. Toyota's strategic managers have determined that about two-thirds of its supply chain, primarily its raw material providers and some of its indirect, parts, and component manufacturers, need to secure ISO 14001 certification. In addition, Toyota is requiring many of these suppliers to phase-out about 450 chemicals and other toxic substances from their manufacturing processes. Another example of sustainable purchasing (previously mentioned in Chapter 6) is the U.S. home improvement retailer, Home Depot. In 1999 Home Depot, the world's largest retailer of wood, began to move away from selling wood products harvested from old growth and other endangered forests in favor of wood products certified as sustainable by the Forest Stewardship Council. Home Depot instituted this sustainable procurement system because of a multiyear campaign by the nonprofit Rainforest Action Network and because of the actions of several dozen of its competitors that had already begun to make the switch to certified wood products.

Issues that purchasing managers may want to consider in their initial approaches to sustainable procurement include: determining whether or not the product or service is really necessary; assessing whether or not the product or service meets the organization's sustainability criteria; evaluating which environmental and human rights standards are being met; examining the potential for reduction, reuse, or recyclability; and assessing the potential of the supplier as a long-term sustainability partner.[17] Of course, there are several barriers to sustainable procurement, including potentially higher costs, excessive time investment, technological issues, and a lack of a sustainable perspective among suppliers. A number of firms have developed responses or strategies to help overcome these barriers, all of which involve developing sustainability-based relationships with suppliers that may include significant sharing of information, costs, benefits, resources, and ideas.[18]

Sustainable Consumption

However, just purchasing sustainable inputs is not enough to make sure the firm's procurement system is sustainable. The second step is to use these supplies within the organization in a sustainably responsible manner.[19] The rate of consumption of supplies within the organization has a great deal to do with organizational members' mental models with respect to sustainability. Thus, educating employees about sustainability can help to reduce the consumption of supplies. As noted in Chapter 2, all firms have two types of flows: matter and energy. Employees make daily decisions involving flows into and out of the organization, ranging from small decisions regarding the use of electricity, paper, or water to larger decisions regarding transportation, manufacturing, and purchasing. In other words, employees are resource gatekeepers. Therefore, for real change to take place, organizational members will need to be willing and able to translate the organization's vision of sustainability into daily practices and decisions.

In office supplies, for example, paper entails a considerable cost to businesses. Organizations engaged in sustainable strategic management might prefer 100 percent recycled content paper, they might select office paper that is chlorine free, or they may require sources that are derived from nontree materials. However, once these sustainable paper products flow into the organization, then employees must integrate sustainable consumption patterns into their daily paper utilization. For example, Dow Europe cut office paper flow by 30 percent in six weeks by reassessing the need for employees to receive written reports and memos.[20] According to estimates by Xerox, if employees increase their duplexing rate on a 60-copies-per-minute copier in a small office, they can realize savings of $1,600 per month.[21] Thus, sustainable procurement not only entails developing sustainability-based supplier relationships and acquiring sustainable input flows, but it also means developing a sustainable consumption culture within the organization to minimize material and energy flows (we will discuss sustainable strategic management cultures in Chapter 9).

Purchasing Services Rather Than Things

An innovative acquisition strategy that is getting significant attention involves purchasing services rather than physical products and materials (discussed in Chapter 6). For example, Nortel in partnership with the nonprofit Pew Charitable Trusts has developed a new type of chemical supply contract. Traditionally, procurement contracts have been based on volume production and pricing, but the new Nortel contracts are not for the chemicals themselves but for the services related to the chemicals. These contracts include agreements for a fixed price per unit, allowing organizations to buy only what is needed and allowing suppliers to produce only what is needed.

This type of contract encourages suppliers to reduce the chemicals they use in the services they provide insofar as it may help them to reduce their costs and increase their profits.[22]

Moving from purchasing physical materials to purchasing services can help to reduce total, average, and marginal use of energy and materials. Further, this transition is generative in nature, creating a new procurement mindset. In this new mindset, for instance, cars may be just one type of personal transportation service, and newsprint may be just one type of communication service delivery system. Firms that can capitalize on this transition from thinking in terms of goods to thinking in terms of services are thus making the kind of fundamental paradigm shifts required for sustainable strategic management.

By creating a sustainable procurement system composed of a sustainable purchasing system, sustainable consumption patterns on the part of organizational members, and innovative procurement contracts, firms can potentially support both low-cost and differentiation SSM strategies. Low-cost strategies may emphasize purchasing items from those suppliers who have lowered their costs through ecoefficiency programs, or they may focus on saving paper in the office or water in the production process. Differentiation strategies, on the other hand, may include advertising that products are produced with organic or recycled materials.

Sustainable Operations Systems

As discussed in Chapter 6, ecoefficient operations systems are the necessary core competence for low-cost SSM strategies, and they are the technologies upon which both manufacturing and service organizations are based. These systems generally engender the greatest organizational investment in physical and human capital. In the biophysical sense, it is via operations systems that organizations transform low-entropy natural resources into high-entropy goods and services. Thus, these systems are critical in the implementation of sustainable strategic management. Recall that ecoefficiency refers to creating and delivering products and services with as little wasted energy and materials as possible by designing environmental efficiency into all operations processes in order to improve operational performance via input-cost savings and the prevention of wastes or low-value byproducts. This means that operations systems are the primary arenas of ecoefficiency in organizations. There are a number of approaches that have been developed to improve ecoefficiency in organizations, and we discuss several of them in this section, beginning with pollution prevention, which is at the heart of ecoefficiency.[23]

Pollution Prevention

Beginning with 3M in the mid-1970s, a number of organizations with substantial manufacturing activities have implemented pollution prevention systems (discussed in Chapter 6) designed to replace traditional pollution control systems, such as scrubbers, filters, deconcentrators, mixers, and so forth. Via such systems, organizations hoped to reduce the costs and impacts of the plethora of strict, expensive pollution-control and reduction regulations. The idea took root quickly because preventing pollution proved in most cases to be easier, more practical, and more cost effective than traditional end-of-pipe pollution-control solutions. Thus, pollution prevention (especially in combination with LCA, DFS, and sustainable procurement) allows organizations to change their operational inputs, processes, and outputs so that fewer wasted byproducts and fewer toxic materials are created and those that are created require less control. This reduces the need for investments in future pollution control equipment and maintenance, and it reduces the impact of regulations on organizations. As noted in Chapter 6, pollution prevention strategies are often depicted as the picking of the low-hanging fruit—that is, they are designed to take advantage of those environmental opportunities with the greatest chance of short-term payoffs.[24]

Among the many organizations that have adopted pollution prevention programs and profited from these are: Republic Engineered Products, which by reducing resources cut its annual costs by $45 million; Baxter Healthcare, which is recycling 99.99 percent of its scrap for $9 million in savings; and Fisher Scientific, a mid-sized company that has implemented twenty-eight pollution prevention projects resulting in $529,000 in annual savings.[25]

Energy Efficiency

Energy efficiency and the substitution of renewable energy sources for traditional ones are important components in creating sustainable operations.[26] More efficient motors, use of insulating materials, and right sizing of pipes and wires, as well as many other techniques, are being employed in operations systems to advance the energy efficiency aspect of sustainability. In addition, organizations now have more opportunities to purchase or produce renewable energy sources and other energy services that do not involve fossil fuel or nuclear energy. Large manufacturers often have the opportunity to use waste heat from their production processes to produce steam for electricity (called cogeneration) that is potentially useful to themselves, their industrial ecology partners, or other utility customers. Firms can also use renewables, especially wind and solar photovoltaic cells. One of the more interesting uses of renewable energy by large businesses is the intention of BP (formerly British Petroleum) to install solar photovoltaic panels on several hundred of its gasoline retail service stations in nearly a dozen countries. The program, projected to cost $50 million, will displace about 3,500 metric tons of carbon dioxide every year, and will utilize panels produced by BP's solar subsidiary, BP Solar (formerly Solarex).

Reverse Distribution

Another technique for moving toward sustainable operations systems is to engage in what is called *reverse distribution*, which involves taking back products from end users when they have finished with them. This approach is gaining significant popularity, especially among some consumer groups, and the need for such programs is currently high. Consider one set of products—electronics. The current wastes in the electronics industry are staggering. Recent reports on the disposal of desktop, mainframe, notebook, and workstation computers, monitors, and peripherals in the United States indicate that recycling rates for these products are 25 percent or less. For instance, of the 1.8 million monitors that consuming organizations say they recycled, only 0.8 million were actually received by smelters for reprocessing. The Sony Corporation has been one of the reverse distribution leaders in the electronics industry, having established a program to recycle Sony batteries in the early 1990s and expanding this effort to other Sony products in the late 1990s. Typically, Sony has agreements with third-party recyclers to handle its recycled products, and it has developed funding schemes to charge customers to recycle products no longer under warranty. Sony claims that this program has reduced its landfill wastes from 120 tons to 5 tons per month.

Total Quality Sustainability Management

Total Quality Sustainability Management (TQSM) is a tool that allows strategic managers to integrate both ecological and social dimensions via a system that links research and development, operations, and marketing in order to improve social and environmental performance. At the present time, social criteria have not been effectively integrated into TQSM. Rather, most firms employ total quality environmental management (TQEM), which focuses primarily on environmental performance. TQEM is an approach for continuously improving the environmental quality of processes and products through the participation of all levels and functions in an organization.[27] TQEM has emerged as one of the most useful and widely implemented frameworks for achieving ecoefficiency because it provides firms with a framework to account for the ecological impacts of products and processes all along the primary type II value chain (see Figure 5.2). By incorporating nature into the total quality management formula, firms are able to achieve high levels of top management commitment and employee involvement in pollution prevention, waste reduction, and so forth. While there are many approaches to TQEM, most attempt to incorporate the sustainability values and perspectives of all decision makers, to use as much qualitative and quantitative information as is available throughout the network, to measure and adjust for increasing levels of environmental effectiveness, and to attempt continuous improvement of the firm's environmental profiles.

One organization that uses TQEM extensively is DuPont. The firm claims that this has allowed it to remain on the leading edge of innovation in cutting wastes and emissions at its plants worldwide. Among DuPont's TQEM successes is the development of new technology used to produce a key raw material for its Lycra product, resulting in the elimination of more than 4 million pounds of wastes and the addition of $4 million per year in revenue. Another TQEM innovation that DuPont has used is process optimization to decrease the firm's emissions of the greenhouse gas HFC-23 by 40 billion pounds, resulting in a savings of $20 million. Such results at DuPont have helped the firm move closer to its overall TQEM goal of zero emissions.

Industrial Ecologies

Although ecoefficient technologies are currently state of the art with regard to sustainable operations systems, McDonough and Braungart (as discussed in Chapter 5) contend that ecoefficiency techniques only make things less bad. Although they note that ecoefficiency is an admirable concept, it works within the same closed circular flow model that has caused humankind's current environmental problems, and merely slows down the process of degradation; ecoefficiency works only to make the destructive system less so. Thus, strategies based on ecoefficiency are transitional strategies in the quest for sustainable strategic management. Eventually, ecoeffectiveness will have to replace ecoefficiency if sustainable strategic management is to be fully implemented.[28]

Many argue that ecoeffectiveness is virtually impossible within a single firm. Instead, they say that ecoeffectiveness can happen only within multifirm industrial ecologies that allow for symbiotic relationships among organizations designed to reduce energy consumption and eliminate pollution and wastes.[29] As previously discussed, industrial ecologies are networks of inter-organizational arrangements where two or more firms attempt to recycle material and energy byproducts to one another as inputs, and so forth. In this approach, multiple organizations are contiguously located in a community. One of the best examples of industrial ecology can be found in Kalundborg, Denmark, in which a coal-fired power plant, a pharmaceutical firm, an oil company, a cement company, a fishery, several farms, and a number of small businesses share industrial inputs and outputs, saving 25 percent or more of numerous resources, including financial resources. As these statistics indicate, Kalundborg (and the other industrial ecologies currently in existence) certainly have not yet reached ecoeffectiveness, but such systems hold a lot of promise for the future in this regard.

As with sustainable research and development and procurement systems, the sustainable operations systems described herein can support either low-cost or market-differentiation SSM competitive level strategies. Pollution prevention, energy efficiency, TQEM, and industrial ecologies are important in developing the core competencies necessary for low-cost SSM strategies, and using renewable energy technologies and TQEM can add environmental features to an organization's products or services, allowing them to charge premium prices or increase market share via environmental differentiation. However, although operations strategies such as those described have effectively integrated ecoefficiency into organizations, they have yet to adequately integrate socioefficiency, and they have yet to reach their socio- and ecoeffectiveness potential. To make the transition to ecoeffectiveness, industrial ecologies will need to become type III industrial ecosystems, TQEM will have to evolve into TQSM, and the other ecoefficient tools discussed above will have to make similar transitions.

Sustainable Infrastructure Systems

Closely related to sustainable operations systems are those infrastructure systems consisting of the physical assets required by organizations. These include the buildings, such as offices, factories, warehouses, parking lots, and other structures owned and operated by the organization, equipment used in and around these buildings, and land and water resources. Buildings often consume 25 percent to 50 percent of all energy used by organizations, and toxins in many buildings can potentially cause human health problems—referred to as the *sick building syndrome*. Thus, an objective for SSM functional level strategies is to target those building areas that are major energy consumers, including heating, air-conditioning, ventilation systems, the building "envelopes" (walls, ceilings/roofs, and flooring), and lighting, and to target the elimination of toxins (black mold, etc.) associated with the sick building syndrome.

Energy audits have been widely available for more than two decades in most developed countries, and the U.S. Environmental Protection Agency Energy Star program certifies new buildings and lighting systems that meet sustainable energy standards. Attention to both current buildings and those yet to be constructed can save organizations one-third or more of their energy costs, often through measures that have payback periods of two years or less. Furthermore, organizations that initiate renewable energy efforts by signing green power contracts or by developing onsite renewable energy systems (such as solar photovoltaic cells) are finding such efforts to be economically viable. Also, the culprits associated with the sick building syndrome can be identified via onsite environmental audits and mitigated through material replacement.

One of the keys to avoiding problems like the sick building syndrome and excessive energy consumption is to design buildings to mimic natural processes.[30] One example of such a building is the International Netherlands Group Bank headquarters, which employs passive solar heating, daylighting, and rainwater recycling. Another example is Ford Motor Company's totally renovated Rouge plant (which we will cover more thoroughly in Chapter 11). Among several benefits of such buildings are energy and water cost savings, lower waste disposal costs, improved heating and cooling systems, and improved employee productivity and health.

Another facilities-related system that offers opportunities for energy efficiency and toxic reductions is the equipment system, whether the equipment is used in or around the office, factory, warehouse, or other building structures. Once again, programs such as the U.S. Environmental Protection Agency's Energy Star program can provide guidance on energy savings regarding new purchases of many types of office equipment, audiovisual equipment, and other electronic equipment. Because most equipment is not yet made with replacement in mind, sustainable strategic managers face tough choices regarding whether and when to replace inefficient equipment. It helps with these choices when several sustainability benefits can be gained from the replacement. For example, replacing equipment to save energy and reduce toxicity becomes an easier decision because such decisions will also reduce the risks of negative present and future employee health effects, thus enhancing the human capital of the firm. Factory equipment might need special attention in this regard because many larger factory machines still employ hazardous chemical-based lubricants and coolants requiring special handling and offsite disposal. Equipment and office cleaning supplies can also cause human health hazards, and thus need to be included in equipment environmental audits.

One final set of infrastructure-related systems that have a sustainability connection are the land and water resources owned and operated by organizations. These can range from large parks and uninhabited tracts of land to vacant lots and storefront surroundings. One issue involves decisions regarding the locations of plant and office sites. To the degree possible, these should be located in proximity to mass transit in order to reduce travel time for employees and fossil fuel consumption. Another land management issue is the protection and, if necessary, restoration of properties that the firm owns. Obviously, toxic waste, hazardous waste, and brownfield sites need to be managed in strict compliance with all appropriate regulations, and adequate funding and waste management practices need to be encouraged. Given that most industrial wastes are stored in nonmunicipal landfills, this is an often overlooked area that could require significant attention from sustainable strategic managers. Of course, the same holds true for onsite wastewater treatment facilities and for all areas where potentially harmful substances are stored. On the positive side, organizations that have extensive undeveloped land and water resource holdings can help ensure that wildlife and indigenous plants and other species are protected on their properties. They may even allow people to visit and enjoy these natural assets in a responsible manner, improving the quality of life for employees, the community, and other stakeholders.

One interesting land preservation project that includes several sustainable implementation approaches is the Lower Mississippi Valley Afforestation Outreach Project in the southern United States. This project is a partnership between the World Business Council for Sustainable Development (WBCSD), the U.S. Department of Agriculture, various state environmental agencies, and three forestry firms—Triangle Pacific Corporation, Westvaco Corporation, and the Temple-Inland Company. These organizations are addressing the problem of significant deforestation caused by logging and land clearing and the resulting negative impacts on agricultural runoff, polluted waterways, and losses of regional biodiversity and wildlife. Essentially, the project broadens the criteria of the U.S. Conservation Reserve Program so that more land can be reforested instead of row-crop farmed. Rent is paid to farmers during the fifteen-year planting period before hardwoods on the property can be harvested sustainably. Once the planting period is complete, the farmers earn income from the sale of the sustainably harvested wood.

Like the other systems described thus far in this chapter, sustainable infrastructure can support either cost savings or revenue-enhancing SSM competitive strategies. Energy efficient assets will likely lead to reduced energy bills and a cost advantage, while wildlife, land, and water preservation may be attractive to stakeholders seeking sustainably differentiated products and services.

Sustainable Marketing Systems

As discussed in Chapter 6, sustainable marketing is the necessary core competence for SSM competitive strategies that focus on ecologically and socially differentiating the firm's products and services in carefully segmented markets. Recall from Chapter 6 that strategic environmental management (SEM) strategies are designed to provide firms only with competitive advantages via ecological differentiation, but SSM competitive strategies expand the scope of traditional SEM strategies to include the social as well as ecological dimensions of marketing. Fuller defines sustainable marketing as "the process of planning, implementing, and controlling the development, pricing, promotion, and distribution of products in a manner that satisfies the three following criteria: (1) customer needs are met, (2) organizational goals are attained, and (3) the processes are compatible with ecosystems."[31]

During the 1990s, the concept of environmental marketing emerged, referring to activities that put product stewardship at the center of the organization's marketing efforts. Environmental marketing has two important objectives: to develop environmentally friendly ("green") products that effectively balance performance, price, convenience, and environmental compatibility; and to project an image to consumers that these products are both high quality and environmentally sensitive.[32] Others ascribe an even higher calling for environmental marketing. They contend that it is an essential prerequisite for transforming the consumer society into a sustainable society.[33]

There are several challenges for environmental marketing. The first is to define the term "green" with all of its implied complexities, methodological issues, and pitfalls. The second challenge is to find ways to convince consumers about the need for lifestyle changes that emphasize greener behaviors and products. Third, environmental marketing is challenged to focus on educating consumers about the true nature of environmental problems and how they relate to the products they consume. Fourth, environmental marketers must find legitimate, nonmisleading ways to communicate the environmental features and benefits of products. The fifth challenge for environmental marketing is to gain credibility for the idea that business interests do not necessarily conflict with environmental responsibility.[34]

The degree of value added related to environmental marketing depends on the interaction between the price or opportunity costs of the product and the environmental benefits of the product. If a green product provides environmental benefits at a lower price or opportunity cost to consumers, then the added value occurs because the green product offers the firm both traditional competitive advantages and environmental advantages. If, as so often happens, a green product is recognized by the consumer to be environmentally beneficial, but the product is more expensive, harder to find, or both, then the value added is achieved only through the product's greenness. If the greenness of a product is not perceived to be at least marginally environmentally beneficial to the consumer and the product's price/opportunity costs exceed those of competing nongreen products, then value added can be achieved only by changing consumer attitudes or providing regulatory structures to support the green product.[35]

Recently, there has been a move to incorporate broader sustainability criteria into the marketing system, constituting a shift from environmental marketing to sustainable marketing. For example, as discussed in the sustainable procurement section above, many firms are now thinking in terms of selling services as replacements for, or adjuncts to, material products, especially in Europe. Ottman has identified four different groups of services that have the potential to substitute for products: (1) product life extension services, which are designed to extend the life of the product via technical assistance, maintenance, disposal service, and so forth; (2) product use services, which entail sharing products and other ways of using products without the need to buy, such as the Greenwheels car-sharing service in the Netherlands; (3) intangible services, which are designed to substitute products for labor-based services such as automated bill paying or voice mail; and (4) result services, which are aimed at reducing the need for material products, such as pedestrian access instead of cars. Selling services has the potential to both lock in customers and increase repeat business.[36] Furthermore, as discussed in Chapter 6, selling services instead of products is an innovative SSM strategy that can be utilized to address the issues of waste and overconsumption in the developed world.

Of course, adding value via sustainable marketing ultimately means developing a sustainable marketing strategy. Key to this is segmenting the market according to customer groups and needs and then developing competitive SSM strategies that meet those needs. As noted in Chapter 5, there are basically three broad segments of environmentally and socially aware consumers: those who are totally committed, those who are committed when the product is price competitive and convenient, and those who do not purchase such products and services. Also mentioned in Chapter 6, environmentally and socially aware consumers tend to be better educated and wealthier. More precisely, research indicates that the segment of the market most receptive to sustainable marketing appeals is composed of educated women, ages thirty to forty-four, with $30,000-plus household incomes. The evidence suggests that women are motivated to keep their loved ones safe from harm, to provide for their children's future, and to be active in social and environmental causes within their communities. In other words, women seem to place a higher importance on environmental and social purchasing criteria than men, demonstrating a greater maternal compassion for the health and welfare of future generations. This segment's potential to influence its peers, and also its purchasing power, make it a very desirable sustainable marketing target.[37]

After carefully segmenting the market, sustainable marketing strategies should be formulated to establish and exploit the firm's social and ecological core competencies. Only through an effective marketing strategy can the firm successfully implement SSM competitive strategies based on social and ecological differentiation. Sustainable marketing strategies should incorporate elements to ensure that consumer needs are met, that the firm's sustainability goals and objectives are achieved, and that the marketing processes are sustainable.

Successful sustainable marketing strategies should include several components. First, they should be based on a clear understanding of the environmental, social, economic, and political issues that affect the organization. Second, sustainable marketing strategies should be supported by strong organizational commitment to sustainability, beginning with top management and permeating all levels of the firm. Third, product and package design strategies should use nature as a framework to develop innovative ways to meet consumer needs in a sustainable manner. Ottman has incorporated the principles of nature into a creative process to generate concepts for new products, services, and packaging that represent minimal social and environmental impacts. She calls this the *getting to zero* process, where the key to packaging is source reduction, and the key to design is to mimic nature by isolating the biological and technical nutrients within the design phase so that the firm is able to utilize a cradle-to-cradle approach as represented by the type III closed-loop value chain.[38]

Fourth, sustainable marketing promotional strategies should be completely honest and nondeceptive; firms should underpromise and overdeliver on their promotional claims. Deceptive sustainability claims can easily trigger charges of *greenwashing* (i.e., making false or misleading environmental and social claims). In addition to the ethical problems of greenwashing, the reputation damage to organizations so charged can be difficult and costly to alter after the fact. A recent trend designed to help firms increase their credibility by promoting their social responsibility is social cause advertising, where firms are integrating social issues into their marketing strategies and using social cause advertising as a means to position their brands for the future. Firms such as Avon, Target, Timberland, and ConAgra have integrated their social commitments as a core component of their marketing strategies in an attempt to enhance their reputations, brand loyalties, and organizational identities.

According to research, corporate investment in social cause marketing promotions jumped more than 400 percent—from $125 million to $545 million—between 1990 and1998. The evidence is that consumers are rewarding those firms who have socially differentiated their products and services.[39]

Another useful tool in sustainable promotion is the use of ecological and social labeling as a means for achieving environmental and social differentiation. As discussed in Chapter 6, ecolabeling is a marketing strategy that is currently being used worldwide to inform consumers about products from construction materials and household appliances to paints and paper products. Certification labels are of three types: those that indicate certification by reputable third parties, those that are self-declared by manufacturers or service providers without supporting evidence, and those that are self-declared with supporting LCA analysis or direct product/service comparisons. The best known of these are in the first category, such as Germany's Blue Angel, the United States's Green Seal, and Canada's Environmental Choice labels. In fact, there are nearly thirty such certification programs, and this has led to a backlash of sorts from some industry associations claiming that these certifications constitute unfair trade barriers. These charges have led to efforts to harmonize these labeling programs through the Global Ecolabeling Network (GEN). In addition, the ISO 14020 environmental labeling guidelines have been developed to attempt to remove much of the confusion resulting from poorly defined environmental marketing regulations.[40]

While many environmental labeling efforts focus on products, some attention is now being focused on the environmental impact of services. For example, the Nordic Swan ecolabel, a joint effort of Denmark, Finland, Iceland, Norway, and Sweden, includes environmental criteria for hotels, covering both energy and toxic material categories. One of the newer environmental certification programs is the Green Globe 21, launched by the World Travel and Tourism Council (WTTC) to establish environmental standards for travel companies and communities. The program, which aims to protect both local cultural and natural resources, certifies travel operations based on both United Nations Agenda 21 guidelines and ISO 14001 processes.

Although not as developed as ecolabeling systems, social labeling, as discussed in Chapter 6, is emerging as a technique for firms to socially differentiate their products and services. Social labeling provides consumers with assurance about the social and ethical impacts of the organization's processes and products. The Ethical Trading Initiative, the Fairtrade Foundation, and the Clean Clothes Campaign all aim to assure consumers that the conditions within a firm's value chain meet basic standards and do not use child labor, bonded labor, or sweatshops in the production of their products. Social labeling criteria are usually based on the conventions of the International Labour Organization (ILO).

The fifth component of sustainable marketing strategies is to ensure that the firm develops the strategies in close consultation with external stakeholders, including regulators, consumers, employees, the media, environmental interest groups, suppliers, and retailers. As noted in Chapter 5, the expanded scope of sustainable strategic management results in a plethora of diverse stakeholder groups with a plethora of interests. Organizations need to engage their network partners in dialogue because input from these stakeholder networks can enhance the quality of a firm's sustainable marketing strategy. An emphasis should also be placed on sustainability education, especially for suppliers, retailers, and consumers.[41]

The sixth sustainable marketing strategy component is pricing. Every effort should be made in the pricing strategy to incorporate the true ecological and social costs into the price of products and services.[42] The pricing component in a sustainable marketing strategy provides the consumer with an indicator of value. Thus, the price should reflect the true costs of resources and all other ecological and social costs of doing business. Consequently, implementing sustainable pricing requires the use of full cost accounting systems (to be discussed further in Chapter 10).

Finally, the seventh component of a successful sustainable marketing strategy is the minimization of the ecological and social impacts of the marketing channels that the firm utilizes. Channel decisions involve both logistics and channel management. With respect to logistics, the focus is on reducing the volume of wastes involved in the flow of products and services. This may involve the minimization of transportation, more efficient inventory programming, and vigilance against accidental discharges. Channel management involves examining the role of retailing, channel partner selection, and various implementation issues associated with sustainable channel decisions.[43]

In sum, sustainable marketing is a critical core competence for organizations wishing to implement SSM competitive level strategies based on social and ecological differentiation. The necessary requirements to achieve this differentiation are: top management commitment to sustainability, the use of LCA and natural processes to design products and packaging, the use of full-cost accounting to develop sustainable pricing strategies, the development of promotional strategies that do not exaggerate social and environmental claims, and an assurance that the channel networks are sustainable.

Conclusions

In this chapter we have described and highlighted the functional level strategies and systems associated with the primary activities of the type III closed-loop value chain. We have examined the sustainability aspects of organizational research and development, procurement, operations, facilities, and marketing systems. Sustainable approaches in each of these systems typically involve identifying the relevant major system components—inputs, processes, outputs, and feedbacks—all along the type III closed-loop value chain, with a focus on improving socio- and ecoefficiency and socio- and ecoeffectiveness. By utilizing the SSM functional level strategies and systems discussed in this chapter, organizations can successfully implement SSM competitive level cost leadership and product differentiation strategies. Successful implementation enhances the firm's ability to achieve its vision of a sustainable firm in a sustainable world.

9

Sustainable Cultures, Structures, Human Resources, and Technologies

In this chapter, we cover some of the sustainable strategic management (SSM) functional level strategies and systems that are critical to the effective support of primary-value-chain SSM strategies such as those discussed in the previous chapter. One of the consistent themes in Parts I and II of this book has been that successful sustainable strategic management efforts require organizational cultures built on sustainability-centered value systems, and that they require learning structures that support these cultures. Further, successful SSM would be difficult if it were not supported by sustainable human resource management systems and sustainable technologies. In a sense, SSM cultures, structures, human resources, and technologies can breathe life into SSM strategies, providing and promoting their human initiation, drive, and renewal, and hope for their success over the long term.

Sustainability-Centered Organizational Cultures

As we discussed in some depth in Chapters 1 and 3, sustainable strategic management cultures are those in which organizational members share artifacts, norms, values, beliefs, assumptions, attitudes, and practices that are consistent with the tenets of sustainability. Recall that strategic managers leading SSM efforts must serve as change agents, guiding their firms through transformational processes in order to fundamentally shift their organizations' cultures toward embracing the underlying assumptions and values of sustainability.[1] In this regard, we presented a discussion of sustainability-centered assumptions and values, and we discussed how these assumptions and values ultimately guide the strategic choices of sustainable strategic managers regarding their organizations' sustainability performance. In this section we discuss some practical ways that sustainability-centered organizational cultures can be developed and maintained.

One key to developing and maintaining sustainability-centered organizational cultures is to develop myths, legends, and stories that highlight and reinforce organizational members' attitudes and beliefs about sustainability.[2] To do this, sustainable strategic managers should foster stories and legends that keep alive the sustainability intentions and challenges of organizational founders and other organizational sustainability heroes. For example, the story of Yvon Chouinard and the beginnings of the outdoor-clothing retailer he founded, Patagonia, is certainly an important story that supports the firm's sustainability culture. Chouinard was an avid mountain climber who initially owned a climbing-equipment company. Among his products were climber's pitons, spike-like objects that are pounded into rocks for climber safety and are nonremovable from the rock once inserted. Chouinard became concerned about the impact the pitons were having on the world's climbing faces and began to shift his company from pitons to chocks, which perform the same function as pitons but are removable from the rock. Chocks became almost as successful a product as pitons and inspired Chouinard to ensure that Patagonia would live by his sustainability-centered values.[3]

Another approach to building sustainability-centered cultures in organizations is to bring community leaders and elders into organizational meetings and other gatherings to relate the history and culture of the local area and how it has changed. In addition managers and employees can relate their own past histories regarding their journeys to sustainability and how they have attempted to advance sustainability during their employment with the organization. Other organizational culture aspects, such as symbols, events, and work (or off-work) routines can be used to support sustainability cultures, such as displaying sustainability slogans (i.e., "think globally, act locally"), giving companywide employee sustainable performance awards, having Earth Day celebrations, and having office "paper out" recycling days. Bringing the outdoors inside, for example, by ensuring adequate daylighting, ventilation, and plants, and bringing insiders outdoors, for example, by sending employees to Outward Bound–type wilderness retreats, are two useful ways organizations can focus their respective cultures on sustainability ideas and activities.

Another effective means to help integrate sustainability into organizational cultures is for organizations to pursue public environmental recognition awards. Dozens of organizational awards are available in most developed countries, many publicized by the environmental ministries of these nations. In the United States, the Environmental Protection Agency not only publicizes such awards programs but also sponsors some, including a Climate

Protection Award, a Small Business Award, and a Presidential Green Chemistry Challenge Award. Others are promoted (or sponsored) by various environmental nongovernmental organizations (NGOs), including the Conservation Fund, the Ecological Society of America, and the Environmental Law Institute.

Approaches like these may have a double benefit. Some organizations, such as Compaq Computer, have seen significant productivity increases as a result of these efforts. Patagonia, mentioned above, describes itself as having a "dirtbag culture" that emphasizes the outdoors (its Ventura, California, location was selected for its proximity to good surfing sites), roughing it, and a comfort level with iconoclasm. Over the years, the company has made many donations to environmental organizations in the form of cash, office space, free products, and employee time, and it has encouraged employees to create recycling and composting systems, adopt energy efficient technologies and practices, and suggest products that are organic or have recycled content.

Visible companywide environmental action programs can also foster sustainability-centered cultures. For example, Sony has developed a series of companywide environmental action programs, the most recent being "Green Management 2005." The program sets goals and implementation strategies to motivate employees at all levels of the company. The program is designed to spur employees in all Sony departments to develop and advance environmental perspectives in their activities, to participate in environmental workshops, to join conservation committees, and to attend annual conferences that help to increase their environmental information, skills, knowledge, and motivation. The firm also uses touring environmental education exhibitions, newsletters, posters, and both intranet and Internet sites to foster environmental innovations, such as the development of lead-free solder and halogen-free printed circuit boards.

One outstanding example from Australia illustrates several sustainability-centered culture-building concepts. Western Power is the major electric utility in Western Australia, and in the mid-1990s it formed a partnership with environmental NGO Landcare and state governments to form the "Green Challenge," a major effort to address one of the region's most important environmental issues—the spread of salinity in agricultural lands leading to severe land degradation. Western Power agreed to provide the coalition with several different kinds of resources, including volunteer accommodations, catering, tree seedlings, and many volunteers from among its employees and their families and friends. The project's objective was to plant one million native trees that were salt tolerant and had antiwater-logging capabilities by the year 2000. To accomplish this goal, Western and its partners used a combination of tried-and-true social cause marketing techniques. They developed an advertising campaign for volunteers that focused on outdoor fun, they sent training information packets, pledge certificates, event planning details, and several reminders to all volunteers and would-be volunteers (including their own employees), and they had the planting sites prepped and ready to go for each of the program's four volunteer weekends per year. Their efforts certainly paid off. They achieved their goal of planting one million trees by 1999, and by the time the project ended, they had planted over 2.5 million seedlings. Organizationally, their employees (and retirees) had a 70 percent volunteer return rate throughout the project's life and spawned a number of Western Power "green teams." For its efforts and results, the company and the project received four major environmental awards symbolizing Western Power's environmentally responsible culture.

One important reason for developing and fostering sustainability-centered organizational cultures is to influence the attitudes of organizational stakeholders over time. It is hoped that helping stakeholders to become more sustainability oriented will lead to greater and deeper levels of sustainable strategic management culture both in the organization itself and throughout its influence network. Regarding the latter, one of the ancillary benefits of creating sustainability-centered organizational cultures is the potential to influence the wider societal culture, especially for organizations that have the size, clout, or standing to take an effective sustainability leadership position.

Sustainability-Centered Organizational Structures

As mentioned in Chapters 1 and 3, it is important that organizations pursuing sustainable strategic management develop effective learning structures that allow them to understand their interconnectedness with their environment, allow them to regularly examine and question their underlying values and assumptions, and allow them, if necessary, to change their values and assumptions. Thus, one of the keys to successful implementation of sustainable strategic management is the creation of these generative learning structures, which are flat, flexible, dynamic, process oriented, and rely on informal, knowledge-based, idea-driven, decision-making processes.[4]

It is likely that the first person who actually recommended an organizational structure appropriate for firms pursuing sustainable strategic management was E.F. Schumacher.[5] He believed that large bureaucracies inhibit freedom, creativity, and human dignity, and damage the ecosystem, and that the only way to reverse the negative effects of vast organizations "is to achieve smallness within the large organization." On this premise, he based his theory of large-scale organization. The theory consists of five principles. The first is that large organizations should be divided into *quasi-firms*, which are small autonomous teams designed to foster high levels of entrepreneurial spirit. The second principle is that accountability of the quasi-firms to higher management should be based on a few items related to profitability, social performance, and environmental performance. The team members make decisions in ad-hoc fashion without interference from upper management; upper management steps in only if the quasi-firm's goals are not being met. Third, the quasi-firms should maintain their own economic identity, they should be allowed to have their own names and keep their own records, and their financial performance should not be merged with other units. Fourth, motivation for lower-level workers can be achieved only if the job is intellectually and spiritually fulfilling and provides ample opportunities to participate in decisions—what Schumacher calls *good work*. According to Schumacher, good work can be achieved in two ways: one method bases the organization's structure on small, autonomous work teams, and the other allows for more employee ownership and participation at the strategic level. The fifth principle of the theory of large-scale organization, the principle of the middle axiom, is that top management can transcend the divergent problem of balancing the need for employee freedom with the need for organizational control by setting broad, strategic directions and allowing the quasi-firms to make their own decisions within these broad directions. Schumacher held that firms organized around these principles would be structured "like nature with little cells," that such organizations resemble a helium balloon vendor at a carnival with a large number of balloons for sale. The vendor (who represents top management) holds the balloons from below rather than lording over them from above. Each balloon represents an autonomous unit that shifts and sways on its own within the broad limits defined by the vendor.

Many now echo Schumacher's sentiments. There is a growing consensus that achieving sustainable strategic management is more likely to occur in organizations that are driven by sustainability-based visions with flat, informal, team-based, knowledge-based, idea-driven learning structures.[6] Learning structures are holistic, interconnected structures that can successfully define their own futures through generative learning processes that allow organizational members to question and change the foundation visions, values, and assumptions that underlie the way they think and behave. These learning structures support the SSM portfolio processes of entrepreneurial learning and dialogue that are critical in formulating SSM corporate level strategies that are based on global sustainability. Learning organizations require vision-driven leadership that focuses on designing, teaching, and stewarding the organization along the path to the vision.[7]

Peter Senge describes five interrelated *learning disciplines* that comprise the framework of learning organizations. The central discipline is systems thinking, which focuses on identifying points of leverage within circular, archetypical models that represent the long-term mutually causal interrelationships among key variables. The other four disciplines include: personal mastery, allowing for the pursuit of personal visions within the context of the organization; a shared organizational vision, which collectively and synergistically represents the employees' personal visions; a willingness to question mental models, that is, to examine basic values and assumptions that underlie organizational actions; and team learning processes, which emphasize generative learning and continuous, honest reflection via dialogue designed to bring the firm's underlying values and assumptions to the surface as legitimate points of discussion and change.

Learning organizations are designed to provide a framework for a more intrinsic, spiritual view of organizational work, organizational life, and organizational purpose. Peter Senge says that in learning organizations there is "a more sacred view of work," and that allowing employees the opportunity to be creative and self-directed in the pursuit of their personal visions is the "spiritual foundation of learning organizations." That is, learning organizations bring a holistic, spiritual dimension into the psyche of organizations, they help create processes that allow for increased meaningful employee participation in strategic decisions, and by doing so they enhance the human capital of the firm. Thus, learning organizations are ideal structures for integrating sustainability into organizational values and visions, and they are ideal structures for facilitating the translation of these sustainability-based values and visions into strategic actions that are in concert with the economic success of the firm and the viability of the greater social system and ecological system.

Sustainable Human Resource Management Systems

Senge's comment that learning persons seeking personal visions are the "spiritual foundation of the learning organization" speaks volumes about the critical role of human resource management in successfully implementing sustainable strategic management. As discussed in Chapter 6, one of the prerequisites for the paradigm shift from strategic environmental management to sustainable strategic management is to view employees as having both instrumental and intrinsic value. Simply stated, without qualified, motivated, dedicated, well-trained, and well-led employees, sustainable strategic management cannot succeed. A number of authors have delved into the human resource requirements, practices, processes, and programs related to sustainable strategic management, and they have identified six highly interrelated human resource factors that are critical to its successful implementation: (1) active support and involvement of top management in the transition to and long-term maintenance of sustainable strategic management; (2) team structures that empower employees to make decisions, generate ideas, experiment, and innovate in order to effectively implement SSM strategies; (3) effective processes and programs for recruiting, selecting, and retaining employees that will add value to the firm's pursuit of sustainable strategic management; (4) effective training, development, and continuous learning programs and processes that maintain the high levels of sustainability knowledge and skills necessary for sustainable strategic management; (5) performance appraisal systems that measure employee contributions to the firm's social, ecological, and economic performance criteria; and (6) reward systems that compensate and recognize employee contributions to the firm's social, ecological, and economic performance criteria.[8]

The first human resource management factor, the need for active top management support and involvement in implementing sustainable strategic management, is a primary theme that permeates this entire book. From the beginning we have emphasized that sustainable strategic managers, especially CEOs, are the primary change agents responsible for instituting sustainable strategic management in their organizations, and, collectively, they are the change agents most responsible for leading the paradigm shift from a closed circular-flow economy to an open living-system economy. We have said that fulfilling these generative change agent roles will require that sustainable strategic managers be effective leaders, who are able to mold sustainability-centered organizational cultures and structures, to meet the needs of the earth's stakeholders, to instill spiritual as well as emotional and intellectual intelligence into the learning processes of their organizations, and so forth.

We have also dedicated significant space throughout the book to the second human resource management factor, the need for team structures that allow for autonomy, innovation, and so forth. In fact, we just discussed in the previous section that organizations pursuing sustainable strategic management need to develop structures composed of learning teams that have the knowledge, skills, and freedom to engage in dialogue, question values and assumptions, generate and experiment with new ideas, build on their sustainability successes, and learn from their sustainability mistakes.

With regard to the third human resource management factor, effective sustainability-based recruitment, selection, and retention, the results of research appear favorable for implementing sustainable strategic management. Attracting, hiring, and keeping talented people in firms seeking sustainable strategic management will be significantly influenced by present and potential employees' perceptions that the organization's vision, mission, and work are socially and environmentally as well as economically responsible. McKinsey researchers found that top talent today wants one or more of four things from their jobs: winning companies, companies that provide exceptional rewards for exceptional performance, companies that provide a high quality of life, and companies that are socially and environmentally responsible (those that can "save the world"). Other research has shown that most college students today say that they want the companies they work for to be socially and environmentally responsible, and that this will be a factor in the job choices they ultimately make. Data have also shown that feeling one's work is important and that it contributes to the greater good improves employee morale and serves as a major factor in employees' decisions to stay or leave a company when other opportunities arise.[9]

It is important to note that successful recruitment, selection, and retention assumes that talented, qualified potential employees are available. In fact, because sustainability is a rather new field with transdisciplinary roots, the pool of talented job candidates with sustainability credentials—such as those with degrees or experience in environmental science, ecological and environmental economics, environmental design, virtual reality, environmental engineering, social development, expert systems, sustainable community development, geographic information systems, safety engineering, sustainable marketing, sustainable accounting, environmental law, sustainable forestry, environmental information systems, sustainable agriculture, and so forth—is not very large and is dispersed across many disciplines and fields. Thus, identifying qualified personnel can be difficult. Organizations may want to search networks and directories such as the National Association of Environmental Professionals or the National Association for Environmental Management. They may also want to advertise in the magazines, journals, and newsletters, and on e-mail list servers and Web sites of various social and environmental organizations.

The fourth human resource management factor required for the successful implementation of sustainable strategic management is the provision of sustainability-based training and development. Because it is important for CEOs and other strategic managers to support and actively involve themselves in successful sustainable strategic management efforts, sustainability training and development should start at the top. In research done in Canadian manufacturing firms, the most common type of sustainability training being provided was environmental awareness training, which was given to 73 percent of the plant managers, 78 percent of the other managers, 76 percent of the salaried employees, and 54 percent of the hourly employees in the study.[10] Apple Computer provides one example of environmental awareness training in organizations. In its program, Apple provides all employees with an environmental, health, and safety handbook, holds regular environmental brownbag sessions, includes environmental items in the company newspaper and intranet, and sponsors a number of Apple Earthday programs.

Because of the complexity of sustainability issues, training and development must go beyond awareness training for many employees. Other important topics for sustainability training include specific social and environmental issues, liability issues and concerns, and public image awareness and management. In addition, extensive environmental training and development covering numerous topics and processes is required for ISO 14001 certification. Instituting total quality sustainability management and design for sustainability also require significant employee training, much of it continuous and ongoing.[11]

The Institute of Environmental Management and Assessment in Scotland suggests that environmental training is important because it can motivate employees to be environmentally responsible on the job by providing sufficient information regarding the prevention of environmental errors and by covering the steps necessary to prevent these errors from becoming disasters. The institute also recommends more informal software-based environmental training approaches in addition to formal classroom strategies. In fact, training and development activities can include traditional in-class or on-line sustainability courses, and it can include wilderness experiences and other external training and development opportunities. For example, organizations may wish to encourage their employees to join an Earthwatch project that combines elements of outdoor recreation, education, philanthropy, and service. It is also important that organizations perform continuous knowledge and skill audits, and they tailor their training and development activities accordingly. These audits help to ensure that organizations maintain the appropriate breadth and depth of sustainability skills.

The fifth human resource management factor is the need for sustainability-based performance appraisal systems. For successful sustainable strategic management to occur, organizational performance appraisal systems should include factors related to employees' sustainability performance. Attainment of social, environmental, and economic goals should be integrated into a holistic performance appraisal system, and these factors should be measured in ways that reveal both short-term and long-term goal attainment. As important as such performance appraisal systems are for achieving sustainable strategic management, research indicates that most organizations are a long way from instituting them.[12]

The final human resource factor for effectively implementing sustainable strategic management is the development of appropriate compensation and reward systems to support its implementation and continuation. These systems may include traditional salary, bonus, or other monetary incentives, or they may include recognition awards such as plaques, cash prizes, verbal praise, and so forth.[13] Of course, a key to effective sustainable strategic management reward systems is that they actually recognize and reward expected sustainability behavior; thus, it is critical that they be used in close conjunction with the sustainability-oriented performance appraisal systems discussed above. One idea befitting sustainably managed organizations is to provide awards that include outdoor recreation packages and the time off to enjoy them. Among the large firms that have incorporated such employee awards into their environmental programs are Baxter Healthcare, Dow Chemical Company, 3M, ICI, and Waste Management.

Sustainable Technologies

As we discussed in Chapter 8, at the core of organizational systems are technologies. Sustainable strategic management technologies come in many forms because organizations are highly varied in their purposes, products, and services. In Chapter 8, we covered some of the ecoefficient technologies associated with the primary value-chain activities, such as pollution prevention, energy efficiency, and so forth. One interesting aspect of these and many other sustainability technologies is that, because the concept and practice of sustainable strategic management is in its infancy and will continue to evolve for some time into the future, the technologies that support sustainable strategic management will also continue to evolve. In this section, we focus our attention on *disruptive technologies* because they present especially intriguing opportunities and challenges to organizations seeking sustainable strategic management in the twenty-first century.

Disruptive technologies are defined as innovative technologies (usually from new firms on the edge) that create dramatic technological paradigm shifts that transform entire industries, economies, societies, or ecosystems. Because of their potential for making type III closed-loop value chains a technological reality, disruptive technologies are considered by many to be at the heart of achieving global sustainability. They are the technologies that, according to economist Joseph Schumpeter, fuel *creative destruction*, which is economic disequilibrium caused by major technological shifts. Pollution prevention and the like are effective ecoefficient technologies that are valuable in creating continuous incremental changes in current products and services designed to serve current customers and markets. However, disruptive technologies hold the promise of providing both socio- and ecoeffective means for developing products and services that can sustainably serve the billions of untapped customers in emerging markets, thus making positive contributions to the economic, social, and natural capital of the developing world.[14]

The discovery of new energy sources throughout human history has resulted in the creation of major disruptive technologies. Think of how changes from reliance on wood to reliance on coal to reliance on oil have dramatically altered the way people live, work, transport themselves, shelter themselves, interact with one another, and so forth. Think what kind of social and ecological changes are possible in the future in both developed and developing markets if humankind is able to shift from fossil fuels to renewable energy sources that are plentiful, inexpensive, and available locally. Solar energy has been and continues to be touted as such a disruptive technology; the cost of photovoltaic cells is going down, the uses for them are expanding, the worldwide market is growing, and the technology is becoming more integrated and efficient. Another energy source with high disruptive potential is the hydrogen fuel cell, especially with regard to transportation. Hydrogen is the universe's most plentiful element, so (like solar energy) no company, country, or cartel can dominate its supply. Furthermore, fuel cells are carbon-neutral systems with zero emissions, so they offer real promise for reducing and eventually eliminating the production of greenhouse gases. In addition, fuel cells not only offer benefits merely from their use but also from their production. Current technology for producing fuel cells relies on natural gas, which does produce carbon emissions, of course; however, there is the potential to produce fuel cells in closed-loop zero-emissions systems using renewable energy sources.[15]

The ongoing electronic/information revolution has certainly provided and will continue to provide disruptive technologies that are changing the way people live, work, and play. There is no doubt that emerging electronic/information technologies have changed the way the world communicates and processes data, and there is no doubt (as discussed in Chapter 6) that there are myriad opportunities for business organizations willing to penetrate emerging markets in developing countries where billions of people thirst to participate in the electronic/information revolution but have not yet had the chance. One revolutionary advance in electronics/information technologies that is particularly useful in sustainable strategic management is the development of geographic information systems (GISs). Since the launching in 1972 of the Landsat Earth-monitoring system, both satellite and ground-level technologies that monitor the planet have increased significantly in sophistication, variety, and number. Together, these systems have evolved into GISs that allow for the development of maps and other graphic images of both natural and human phenomena. Governments, businesses, and NGOs now use GIS data regularly to make more informed environmental and social decisions. With GIS data widely available, humans now have the potential to create much clearer pictures of the earth's observable natural resources and the effects on them of human development, pollution, and depletion. From observing the effects of acid rain and overfishing to determining the best sites for solar and wind energy installations, GISs promise to provide improved quality (and quantity) of social and environmental data to organizations seeking sustainable strategic management into the foreseeable future.

Another fruitful area for the development of disruptive technologies is biotechnology. A particularly controversial arm of biotechnology is the development of recombinant DNA (rDNA) technology—cloning. The dark side of rDNA research and technological development involves fears of massive crop disasters resulting from "runaway genes" and fears of human health problems related to the consumption of genetically modified organisms. On the brighter side, advocates of developing rDNA technologies view it as a part of a green revolution that will see food production increase many times through the development of disease-and pest-resistant plants, and rDNA cloning technology may allow humans to more effectively save endangered species.

As mentioned above, disruptive technologies provide potential avenues to achieving socio- and ecoeffectiveness. For example, the development of technical and biological nutrients (discussed in Chapters 5 and 8) provides many opportunities for ecoeffective strategies. Cargill and Dow are pursuing such a strategy via a joint venture, called Cargill Dow, to develop a line of fiber and packaging materials called NatureWorks that are made completely from corn and other agricultural crops. In the process, the plant's natural sugars are fermented, condensed, purified, and naturally synthesized into a material that competes with chemically based synthetics, such as nylon, polystyrene, and polyethylene, in a wide range of products, including clothing, upholstery, and packaging materials. NatureWorks is a biological nutrient, a product that is completely biodegradable and will fully disaggregate when composted. Of course, this material is not yet completely sustainable because the corn and other agricultural products that comprise it are currently dependent on fossil fuels for the pesticides, herbicides, and gasoline-consuming farming machinery used in their growing, harvesting, and transportation stages.

Many believe that space exploration also provides potential opportunities for the discovery of disruptive technologies. While many believe that spending money to explore outer space drains resources and attention away from protecting earthly treasures and bettering the quality of life of its neediest citizens, others believe that space exploration can provide new perspectives and vital data on the planet's evolving conditions. Nonetheless, numerous technologies developed from space exploration have made their way into everyday life, and the potential for this stream to continue is certainly there. However, the real questions are, how far should humankind's space vision reach? Should humans be satisfied with quality-of-life improvements on earth that technological advances from space research can afford, or should they seriously consider space technology as a means of escape from the planet if and when it becomes unsustainable?

As discussed in some detail in Parts I and II of this book, one of the primary concerns regarding the pursuit of new technologies is whether or not the poorest members of humanity will ever benefit from these technologies. Unless they do, the technological divide between the wealthy and the poor will continue to widen.[16] One of the beauties of disruptive technologies is that they can help nations reverse poverty trends, thereby helping them to increase both the quality of life and age expectancies of their people. Such technologies can help nations experiencing poverty today to leapfrog into the future. Rural communities and villages in developing countries can forgo developing traditional electric generation plants and distribution systems in favor of adopting distributed solar photovoltaic electricity systems. Other communities in developing countries can jump from no telephone and Internet service to wireless versions of both with their lower infrastructure and ecological costs.

In sum, energy transformations, the electronic/information revolution, space technology, biotechnology, and other technological advances have led to and will continue to lead to virtually unlimited opportunities for the discovery of disruptive technologies that can change the way humans live on the planet, making positive contributions to both human and natural capital. Disruptive technologies are key to making the shift from socio- and ecoefficiency to socio- and ecoeffectiveness, and they are key to bringing emerging markets into the economic realm in sustainable ways. It is no wonder that many consider disruptive technologies to be a giant piece in the sustainable strategic management puzzle.

Conclusions

Sustainable cultures, structures, human resources, and technologies are essentially the hearts and minds of organizations pursuing sustainable strategic management. Combined, they set the stage for effective implementation of SSM strategies over the long term. The challenge for sustainable strategic managers is to provide ways to build real strength into each of these elements and to coordinate them so that their organizations can achieve their sustainable strategic management goals and objectives. The organizations that are making huge strides in their efforts to implement sustainable strategic management are those that build their cultures on sustainability-centered value systems; create learning structures that allow them to think generatively; hire, retain, develop, and reward employees with the skills, knowledge, and motivation to contribute to effective sustainable strategic decisions; and develop or take advantage of disruptive technologies.

10

Evaluating Sustainable Strategic Management

Effectively implementing sustainable strategic management (SSM) cannot be done without regularly assessing the degree to which SSM strategies achieve their respective goals and objectives. SSM strategy evaluation begins with a generative evaluation of the fundamental assumptions on which the strategies are based. This entails reviewing, evaluating, and questioning data from both external (Chapter 4) and internal (Chapter 5) analyses. The next step in SSM strategy evaluation is to measure sustainability performance, comparing it against sustainability-based goals and objectives in order to identify performance gaps in the firm's current SSM strategies. This gap analysis reveals the extent of change needed in current SSM strategies in order to achieve the expected sustainability performance standards, thus providing the impetus for corrective actions on the part of sustainable strategic managers. As a matter of practice, sustainable strategic management evaluation occurs organization wide, with all organizational members, regardless of level, continuously questioning whether or not the SSM strategies within their realm of responsibility are achieving their intended results. However, there are three value-chain support activities that are particularly important in evaluating SSM strategies: accounting and finance, information systems, and reporting.[1] In this chapter, we briefly present some of the more prominent measures and external indexes that organizations can use to evaluate their SSM strategy performance, and we discuss these three critical value-chain support activities mentioned above in the SSM strategy evaluation process.

Sustainable Strategic Management Performance Measurement

While measuring economic performance is an age-old process, evaluating social and environmental performance is a new and emerging discipline. As we will see in this section and the sections below, the evaluation standards, measurement criteria, certification systems, information systems, reporting systems, and so forth are all in the neophyte stages of development and implementation. Nonetheless, concepts, criteria, and processes are emerging that can assist organizations in examining their social and environmental performance. Measuring environmental and social performance is critical in successful SSM strategy evaluation for several reasons. First, measurements clarify how strategic managers define sustainability within their firms, and they provide a mechanism to monitor social and environmental progress against specific indicators. Second, developing specific measurements for sustainability performance engages managers at all levels, especially if some of the measurements are consistent with traditional investment measurements. Third, by providing a range of common measurements, they enable clear and consistent sustainability reporting to internal and external stakeholders.[2]

Lately, there has been an effort to develop common environmental performance indicators (EPIs). EPIs include either annual or per-product measures for things such as energy efficiency, expenditures on sustainable management systems (SMSs), stakeholder interactions, air emissions, consumption of paper, water, and energy within the firm, and so forth. As nongovernmental organizations and other third parties benchmark and compare the environmental and social performance of competing firms and rank them accordingly, this provides incentives for firms to incorporate sustainability into their strategies. The mere act of measuring and reporting performance, both good and bad, opens up the firm to public scrutiny and promotes a positive corporate image of a responsible firm.[3] Other related techniques have been developed; for example, the stakeholder value analysis toolkit facilitates stakeholder interactions with respect to the organization's social and environmental performance. Although environmental performance measurements are gaining a wide acceptance, corporate social responsibility measures need more development.[4]

Lober suggests four organizational effectiveness models that may be applied by organizations seeking to evaluate their SSM strategy performance. These models include: the organizational goals model, which focuses on the organization's success in attaining its sustainable strategic management goals; the systems resource model, which focuses on the organization's ability to capture and efficiently use social and environmental resources; the internal programs model, which focuses on the organization's internal environmental and social performance; and the strategic constituencies model, which focuses on managing the organization's socially and environmentally concerned stakeholders.[5]

Regardless of which of these (or other) models firms use to organize their evaluation processes, they will have to select the social and environmental criteria and measurements that are important to their firms' sustainable strategic management performance. At a minimum, organizations should be able to evaluate their regulatory performance, which requires them to keep and report data on toxic emissions, habitat disturbance, pollution, discrimination, waste disposal, fair hiring practices, and so forth, related to the plethora of social and environmental regulations discussed in Chapter 5. However, as we have emphasized many times, sustainable strategic management requires going beyond the regulations; it requires doing much more than the minimum. In fact, there are several performance levels for strategic managers to consider when assessing their sustainability-based performance, ranging from regulatory compliance, to public disclosure, to beyond compliance, to corporate sustainability, to global sustainability.[6] Clearly, strategic managers wishing to guide their firms toward sustainable strategic management will want to aim their firms' performance at the highest level, global sustainability. This is the level at which firms transcend the eco- and socioefficient view of sustainability as a business opportunity (type II value chain) to the eco- and socioeffectiveness view of sustainability that requires embracing global stewardship that includes the developing world (type III value chain). Therefore, organizations pursuing sustainable strategic management will want to measure a wide variety of social and environmental factors for both developed and developing world markets, benchmarking the results of their performance in these areas to other firms seeking sustainable strategic management so that they can continue to improve their sustainability performance.

The factors used to evaluate sustainability performance are as varied as the industries, cultures, and ecosystems affected. Some of the most common measures for environmental performance in manufacturing (as set out by the World Business Council for Sustainable Development [WBCSD]) include: reduction in material intensity, reduction in energy intensity, dispersion of toxic substances, increases in recyclability, the use of renewables, durability, and service intensity. Other criteria used to measure environmental performance may include items such as: habitat incursion and destruction, soil erosion, water resource impact, species vulnerability, greenhouse gas contributions, waste stream management, and so forth. With regard to measuring social performance, there is a wide range of criteria, including: corporate ethics, employee safety, employee rights, employee work life, employee health, consumer rights, product safety, philanthropy, volunteer activities, community relations, child and forced labor, social equity, diversity, discrimination, labor union issues, cultural preservation, and so forth.[7]

External validation of sustainable strategic management performance is considered to be critical to the SSM strategy evaluation processes because it adds credibility to an organization's sustainable strategic management claims. Many external organizations establish and measure numerous sustainability criteria such as those described above. We have already mentioned or will mention later in the book a number of these, including: the WBCSD's seven ecoefficiency criteria, the British Standard Institution's BS 7750, the European Union's Eco-Management and Audit Scheme (EMAS), the International Organization for Standardization (ISO) 14000 process, the International Chamber of Commerce Business Charter for Sustainable Development, the Chemical Manufacturers Association (CMA) Responsible Care guidelines, the Global Environmental Management Initiative programs, the CERES Principles, the Council on Economic Priorities (CEP) guidelines, the International Labour Organization (ILO) conventions, the United Nations Declaration of Human Rights, the Forest Stewardship Council sustainable wood guidelines, the Innovest EcoValue'21 research platform, and the Dow Jones Sustainability Index.

In addition to these, there are a number of other organizations that set and measure social and environmental performance. In the environmental arena, there is the Eco-efficiency Assessment per Unit of Service (ECOPUS), which uses life-cycle analysis (LCA) to help organizations identify areas of concern regarding their performance on the WBCSD's seven ecoefficiency criteria. There is also Environmental Defense's Chemical Scorecard, which investigates chemical pollution sources, concentrations, and severity in the United States, stratifying the data according to zip code, county, state, or industrial facility. The European Chemical Industry Council has instituted guidelines similar to the CMA Responsible Care guidelines for the European chemical industry, and the National Academy of Engineering measures environmental performance in four industries—automotive, chemical, electronics, and pulp and paper.[8]

In the social arena, some of the external organizations that provide evaluation criteria and processes include: Business for Social Responsibility, which provides measurement and assistance to its 1,400 members in areas such as human rights, community development, the environment, corporate governance, and accountability; the Interfaith Center on Corporate Responsibility, which has established its Principles for Global Corporate Responsibility; the Investor Responsibility Research Center, which publishes a regular newsletter reporting on corporate social performance; and the Social Venture Network, which has developed ten principles of corporate social responsibility as well as a process for implementing and measuring these in organizations. In addition to these examples, there are dozens of other organizations providing similar guidelines. The point is that there are many organizations providing a wide variety of criteria, methods, and frameworks that can be valuable in helping organizations evaluate and benchmark the effectiveness of their SSM strategies.

Sustainable Accounting and Finance

Typically, the finance and accounting subsystems in organizations are designed to track financially related transactions and trends that allow organizations to assess past organizational results, to provide information regarding how the organization can secure future financial resources, and to determine how these might be spent or invested. Many aspects of sustainable strategic management, such as the costs of recycled materials, investments in community training programs, installation of pollution prevention and control equipment, and marketing costs related to socially and environmentally responsible products, have financial components that can be recorded and tracked using traditional financial accounting methods. Traditional financial accounting methods may be useful in: discovering when products, services, or production processes are wasteful and environmentally damaging; calculating the additional revenues related to sustainability-oriented products and services; ensuring that capital budgeting processes include social and environmental benefits and costs; and identifying other revenue-enhancing sustainable strategic management opportunities for the organization.[9] The Financial Accounting Standards Board (FASB), in its Issue 90–8, requires that the costs of treating contamination, including the costs of problem analysis, the costs of toxic substance removal and neutralization, and the costs of preventing future contamination, should be charged as expenses to firms; and FASB Opinion Number 5, Accounting for Contingencies, requires that firms record potential environmental losses if they can reasonably assume that these losses will actually occur in the future and that their costs can be reasonably estimated.[10]

However, traditional financial accounting has some basic pitfalls when applied to sustainable strategic management. There are a number of sustainable strategic management issues that, to this point, cannot be properly accounted for using traditional financial accounting methods. For example: What is the economic value of a cubic meter of clean air? How much is the aesthetic beauty of the land worth? What monetary amount can be placed on the psychological costs of human displacement resulting from environmental or social upheaval? How much value can be assigned to future generations of human beings? How much value can be assigned to other species, now and in the future? At the very least, the time scale related to such questions renders traditional financial discounting methods virtually useless.

The inadequacies of traditional financial accounting methods for sustainable strategic management have led to calls for the development of *full-cost accounting*, financial accounting systems that account for the environmental and social as well as economic costs associated with doing business, including the costs of externalities incurred by society and the planet.[11] These systems fully integrate economic, social, and environmental criteria, assign fundamental rather than secondary importance to social and environmental concerns, account for all internal and external costs now and in the future, and reflect long-term financial performance.[12] For example, the full costs of an automobile would include all of the traditional costs associated with designing the car, acquiring the materials, producing and transporting the parts to the manufacturer, assembling the car, and delivering it to the customer, but they would also include the costs of potential air pollution, public health, resource depletion, congestion, injury, death, and blight. As discussed in Chapter 8, sustainability pricing can be implemented only if strategic managers are committed to using full-cost accounting.[13]

One group committed to both research and development of full-cost accounting is the Association of Chartered Certified Accountants (ACCA) in the United Kingdom.[14] The ACCA's commitment to developing full-cost accounting is rooted in the European Commission's Fifth Action Program declaration, which stipulates that the consumption and use of natural resources in products and services should be reflected in their market prices. According to the ACCA, full-cost accounting involves four steps: (1) defining the cost objective, that is, determining which products, processes, strategic business units (SBUs), and so forth are the targets of the analysis; (2) determining the scope of the analysis, including deciding on the key externalities that need to be measured; (3) measuring the external impacts of the product, process, and so forth; and (4) determining the monetary value of these measured external impacts. As would be expected from the above discussion, ACCA researchers have found that the most difficult issue associated with implementing full-cost accounting occurs in step four—assigning monetary value to the externalities.

The ACCA suggests three techniques that firms can use to account for externalities: life-cycle analysis, ecobalances, and ecological footprints. We have already discussed life-cycle analysis in some depth. Ecobalances are systems designed to track energy, resource, and material flows against product/service outputs, emissions, leakages, and wastes for specific business entities—specific products, services, processes, SBUs, and so forth. The ultimate goal of ecobalancing is to balance energy and material inputs with energy and material outputs absent any significant leakages, emissions, and wastes (hence, creating a type III closed loop). Ecological footprints have historically been done for entire countries, and have generally led to revelations about the social and economic inequities related to excessive consumption and wastes in developed nations as compared to developing nations (recall the Netherlands example in Chapter 4). The ACCA says that firms can determine ecological footprints just as well as nations. For example, using ecological-footprint analysis,[15] firms can calculate not only the financial costs of air travel but also the number of miles traveled and resources used per mile for that travel. The ACCA report notes that neither ecobalances nor ecological footprints can be fully expressed in financial terms at this point, but the researchers believe that, for the most part, it is possible to do so, and they recommend continued work in this regard.

The ACCA researchers compared full-cost accounting experiments in five organizations—BSO/Origin, Ontario Hydro, Interface Europe, Anglian Water, and Essex Water. They drew several conclusions from their comparison: (1) Transportation is one of the most significant generators of externalities, but the current economic system significantly restricts firms from making much improvement in this regard. (2) By working to account for environmental externalities, firms learned what they needed to improve and got insights into how to do it. (3) Not all the work in these examples was done retroactively; some of the organizations were accounting for externalities before they occurred. (4) There was a wide variety of costing methods used in the five cases, making comparisons among organizations difficult. However, one commonality found was that, regardless of the costing method used, full costs were often seen as exceeding traditional profit figures when the cost of externalities was added into the formula.

A number of organizations are putting full-cost accounting to use. Governmental agencies across the globe have mandated its use for processes such as municipal solid waste management. For example, New Zealand requires that full-cost accounting be used for all landfill planning, location, development, operation, closure, and aftercare, and the State of Florida requires essentially the same for all municipal landfills.[16] Some firms are working to develop full cost accounting systems, including Dow Chemical and Baxter International. In Baxter's case, the firm calculates and reports its monetary positive and negative environmental impacts as subsets of more traditional accounts, breaking out those sustainability items that can be identified and reported to users.

Full-cost accounting is certainly in its formative stages, and the barriers to its development and universal implementation are significant. As it stands today, there is certainly a wide gulf between full-cost accounting theory and practice. In addition to the problems related to assigning meaningful monetary value to things like aesthetics and future generations, current full-cost accounting practices aim primarily at the environment and focus only slightly or not at all on social costs. Of course, including such costs is critical to the effective long-term implementation of sustainable strategic management.[17] However, if the work of the ACCA is indicative of what is happening with regard to the development of full cost accounting, then we are encouraged that this tool will some day provide a powerful support system for the implementation and evaluation of SSM strategies.

In addition to full cost accounting, there are several financial indexes that allow firms to get a sense regarding the relationship between their financial performance and their social and environmental performance. One of these indexes is the Innovest EcoValue'21 research platform, which assesses more than 1,000 firms on sixty different environmental criteria and provides ratings between AAA and CCC. Innovest works to identify the best-in-class firms, the ones that it predicts will outperform the market over time. Innovest investment advisers collect information in five categories to make this assessment: historic environmental contingent liabilities, environment-related operating risk exposure, environment-related managerial risk efficiency capacity, and strategic environmental profit opportunities. Information on each of these factors is collected, weighted, totaled, and compared among competitors in the same industry set, and from these results assigned Innovest's bond-like triple letter rating.[18]

Another index to connect financial performance with sustainability is the Dow Jones Sustainability Group Index (DJSI). The DJSI is actually a set of indexes related to the financial, social, and environmental information on a selected set of almost 3,000 firms in more than thirty countries. These sustainability profiles include firms in nearly seventy industries, with a combined end-of-decade market capitalization of $5.5 trillion. These indexes include a global index and three regional indexes (including one for the United States). The firms are selected based on environmental and social criteria gleaned from a survey of their top managers and on an assessment of industry-specific trends and forces. This combination of factors allows the DJSI to be used to make projections on the sustainability futures of the companies in the index. (Data indicate that over an eight-year period, the DJSI firms would have outperformed several of the traditional Dow Jones indexes.) Like other financial market indexes, analysts, brokers, and investors worldwide use the DJSI to make financial decisions at the firm, industry, and market levels.[19]

Sustainable Management Information Systems

Effective management and use of information is critical for all stages of sustainable strategic management—formulation, implementation, and evaluation. However, it is in the evaluation stage that data on economic, social, and environmental performance are aggregated, integrated, analyzed, compared to goals and objectives, and reported in usable form to sustainable strategic managers. A sustainable management information system (SMIS) is a management information system designed to identify, collect, process, and transmit environmental and social information from throughout the organization's internal and external environments in order to improve the decisions of sustainable strategic managers. Having reliable internal and external sustainability data is important in the first stage of the evaluation process—examining the fundamental assumptions upon which the strategy is based. An SMIS is used to track information on the organization's environmental and social impacts, such as its air emissions, water effluents, or product safety issues. Collecting and processing such information effectively has a number of organizational benefits, such as allowing easier calculation of full environmental costs throughout various processes, contributing to life-cycle analyses of products and services, assisting in identifying opportunities for research and development, and identifying target markets for socially and environmentally responsible products and services.

Like most information systems, an SMIS includes three elements—hardware (i.e., computers, global positioning systems, etc.), software (i.e., sustainable strategic management applications programs), and warmware (i.e., human designers, operators, and analysts of the system). Because sustainable strategic management data and decisions vary across such a wide spectrum of issues and processes, many sizes and configurations of SMISs are possible. They may range from the supercomputers and sophisticated modeling programs used in climate change projections to simple spreadsheets on a personal computer indicating the appropriate sizing and positioning of solar photovoltaic panels.

With regard to software, there are systems that allow for social and environmental information to be incorporated within traditional management information systems. For example, software is available that measures the firm's environmental performance against ISO 14000 performance evaluation guidelines (ISO 14031), thus enabling strategic managers to evaluate their progress toward ISO 14000 certification.[20] In its early stages, SMIS software was rather complex and expensive, but there are efforts now under way to create software with lower front-end costs and easier initial access and use. For example, the Global Environmental Management Initiative has developed its SD Planner (sustainable development planner) using Microsoft Access 2000, which is a very common and widespread platform. The software can run on most computers using Microsoft Windows, and it is easily customizable. This program guides system operators through several stages of prompts in the formulation of their organization's sustainability plan, and it provides the means for evaluating progress toward the organization's social, economic, and environmental goals and objectives set out in the plan.[21]

SMISs are growing rapidly in both number and sophistication. The market for such systems is expected to reach $6 billion by 2005, and software is now available to monitor and control material mass, process flows, waste management, environmental health and safety, product safety, LCA processes, and so forth.[22] Furthermore, with advancements in remote sensing, processing, dissemination technologies, and so forth, sustainable strategic managers can look forward to better systems that allow them to make even more informed decisions regarding their firm's sustainable strategic management performance in the future.

One of the keys to an effective SMIS (or any information system, for that matter) is to establish priorities for data gathering, processing, and analysis. Generally, this means starting with those data required by regulation. Because of the proliferation of social and environmental regulations since the 1960s (see Chapter 5), organizations are now required to gather, process, and report data on hiring practices, worker health and safety, air pollution, water pollution, waste management, toxic substance control, community right to know, pollution prevention, greenhouse gas production/reduction, chlorofluorocarbon production/reduction, and so on. For each of these categories, organizations are required to create planning documents, permit applications, monitoring systems, accounting systems, disclosure systems, pollution reduction systems, and so forth.[23] To illustrate, large manufacturing firms that generate or transport nonnegligible amounts of any of the hundreds of chemicals and materials listed in the U.S. Environmental Protection Agency's Toxic Release Inventory (TRI) regulations will have to make the manufacture, handling, transportation, and disposal of toxic materials a high SMIS priority. Such a priority is certainly justifiable on grounds that the substance(s) in question is regulated, and it is also justifiable because TRI-listed substances are very dangerous to human health and safety, and their mishandling can cause serious injury and death as well as the potential for huge legal liabilities for organizations.

Sustainable Strategic Management Reporting

It is generally agreed that accurate and full reporting on a firm's social and environmental actions and performance to its stakeholders is a critical sustainable strategic management activity. Sustainability reports should meet several effectiveness criteria, including: covering the material completely in comprehensible ways; responding to stakeholder inquiries and concerns; ensuring both continuity and comparability of data over time; fully describing all items (activities, products, processes, etc.) related to implementing the firm's SSM strategies; presenting an overview of the firm's sustainable strategic management policies, programs, and performance targets; and reporting on both normal operations and unusual events or incidents.[24]

Some reports are mandatory, such as the aforementioned U.S. Environmental Protection Agency's Toxic Release Inventory requirements, which give the public access to vast amounts of environmental performance data. However, many corporations today are taking a more proactive approach to reporting their sustainable strategic management activities and results. In addition to the traditional shareholder reports that focus on economic performance, organizations are creating detailed reports that focus on their social and environmental performance (a number of which have been sources of information in this book). These reports are generally widely distributed to employees, shareholders, financial institutions, customers, local communities, interest groups, the media, regulators, and many times to the public at large (via the Internet, discussed further below). A common practice (and in some cases a requirement) is to use external sustainability indexes such as those discussed in previous sections as the basis for these reports. For example, many corporations use the WCSBD ecoefficiency criteria and the United Nations Declaration of Human Rights as frameworks for reporting their sustainable strategic management performance. According to the United Kingdom's Association of Chartered Certified Accountants, some firms are doing a good job of reporting their social and environmental performance (e.g., BP Amoco and Ben & Jerry's), but others have managed to go a step beyond, providing sustainability reports that effectively integrate economic, social, and environmental performance (i.e., Shell, Procter & Gamble, Interface, and Bristol-Myers Squibb).[25]

One very extensive sustainability reporting effort is the Global Reporting Initiative (GRI). Started by CERES in 1997, GRI became independent in 2002, and it is currently affiliated with the United Nations Environment Programme. The focus of the GRI is the development of environmental, social, and economic reporting guidelines to help advance global comprehensiveness and consistency in sustainable strategic management reporting. In this approach, organizations are asked to report on three types of sustainability criteria: generally applicable indicators, company-specific indicators, and emerging indicators. GRI guidelines not only include separate criteria for environmental, social, and economic performance data, but also ask organizations to report on the interactions among these three. For example, under GRI guidelines, firms may want to report how they measure their environmental justice efforts and outcomes in order to ensure that their environmental activities do not have a disproportionately negative impact on low-income or minority communities.[26]

Another popular sustainable strategic management reporting approach has been developed by SustainAbility, a consulting firm located in the United Kingdom, in partnership with the United Nations Environment Programme. This is a five-stage environmental performance reporting process. Stage one involves reporting on the firm's environmental policies and systems, and stage two involves reporting on the firm's environmental inputs and outputs. Stage three involves reporting the financial results of the firm's social and environmental management efforts, stage four involves reporting on the firm's relationships with its social and environmental stakeholders, and stage five involves reporting on how the firm's sustainable strategic management strategies and processes contribute to overall sustainable development.[27]

One of the most important current trends in sustainable strategic management reporting is to ensure that sustainability reports (and other corporate reports) are themselves produced and delivered in sustainable ways to shareholders and other stakeholders. Thus, sustainability reports are now regularly being provided on compact disc or made available on the Internet.[28] Using electronic media to produce and transmit reports has a number of economic, social, and environmental advantages. Such media save on production and transmission costs, allow for more in-depth information to be distributed to more people from more cultures and social strata across the globe, and save both paper and energy. An excellent example of an in-depth, informative, globally available sustainability report is Ford Motor Company's *Corporate Citizenship Report* (available at www.ford.com). The report reflects CEO and chairman Bill Ford's underlying values and beliefs with regard to the firm's commitment to sustainability, details Ford's SSM strategies, reports on the social and environmental partnerships Ford has joined in its efforts to implement these strategies, discusses its ecoeffectiveness production efforts, reveals its sustainability-based responsibilities to its stakeholders, and so forth.

Before closing this section, we will discuss one firm that does its sustainability reporting a little differently. Seventh Generation Inc. focuses on reporting the positive impacts its products have on the natural environment as compared to its competitors, and it reports this information in its promotional literature to its customers. To do this, the firm calculates the amounts of pollution and resources that can be offset if consumers purchase its products (which are made with nontoxic, renewable, recycled materials) rather than purchasing products from its traditional competitors. The company has found in such evaluations that its products have offset literally tons of various pollutants and prevented tens of thousands of trees from being cut unnecessarily. Once these evaluations are complete, Seventh Generation reports the results in its promotional literature (catalogues,

Web site, packages, etc.). For example, it publicizes on the wrapper of its bathroom tissue that if every U.S. household were to replace just one roll of regular tissue with its 100 percent recycled content tissue, the savings would amount to 423,900 trees, 1 million cubic feet of landfill space, and 153 million gallons of water.[29]

Conclusions

In this chapter we have discussed evaluation processes and systems that support sustainable strategic managers in their efforts to evaluate the effectiveness of their firm's SSM strategies. Collectively, these support activities can provide organizations with guidance regarding where they are and where they want to go regarding their sustainable strategic management goals and objectives. However, a warning is in order. Sustainable strategic management is in its infancy, and the same can be said about the tools used to evaluate its effectiveness. For example, we discussed in the full-cost accounting section that there are a number of critical social and environmental dimensions that simply cannot be evaluated using the traditional financial accounting approaches. Thus, as with formulation and implementation, effective evaluation of sustainable strategic management will eventually require a paradigm shift regarding the way value is assigned to people and to nature.

11

Toward Sustainable Strategic Management

As we come to the end of this book, we would like to point out that over the past two decades or so, there has been significant progress in the transition from traditional strategic management to strategic environmental management (a transition from a type I eco- and socioinefficient system to a type II eco- and socioefficient system as described in previous chapters). Concepts such as pollution prevention, product stewardship, materials recycling, and so forth are becoming ingrained in the creation, manufacturing, and marketing processes of all types of organizations, and tools such as life-cycle analysis and full-cost accounting are growing in both sophistication and use. Such frameworks and tools have proved to provide competitive advantages for organizations, allowing them to improve their environmental performance while, at the same time, improving their economic bottom lines. However, it is important to note that, although practicing strategic environmental management definitely opens organizations to environmental and social concerns, it is in a way that does not challenge the dominance or content of the basic economic paradigm. In strategic environmental management, firms save materials and energy, provide consumers with more environmentally friendly products and services, serve their communities, and reap financial gains in return.

As we have contended throughout the book, while strategic environmental management is a very important development along the journey to sustainability, by itself it is unlikely to be enough. Organizations wishing to positively contribute to a world that can support a high quality of life for their children and grandchildren will need to make the transition from strategic environmental management to sustainable strategic management (from a type II eco- and socioefficient system to a type III eco- and socioeffective system). However, without a shift in the dominant economic paradigm, making this transition will be difficult. Below we return to the stories of Anthony Flaccavento, Bill Ford, and their organizations (introduced in Chapter 1), in order to demonstrate some of the current economic realities related to pursuing sustainable strategic management.

Anthony Flaccavento, Bill Ford, and the Real Triple Bottom Line

We opened this book with the stories of two strategic managers driven by visions of sustainability, Anthony Flaccavento, executive director of Appalachian Sustainable Development, and Bill Ford, chairman and CEO of Ford Motor Company. Recall that both of these executives have dreams of achieving economic success for their organizations via processes that protect and enhance the planet's natural resources and the welfare of its citizens. In Chapter 7 we presented TOWS matrixes of the sustainable strategic management (SSM) strategies these firms have developed to manage their threats and weaknesses and take advantage of their opportunities and strengths. We now revisit these men and their organizations, focusing on the SSM strategies they have formulated and implemented, the progress they have made, and the roadblocks they face as they pursue sustainable strategic management in the future.

Anthony Flaccavento and Appalachian Sustainable Development

As mentioned in Chapter 1, Anthony Flaccavento and his colleagues started Appalachian Sustainable Development (ASD) in 1995.[1] Anthony's upbringing, university studies, and experience became the sources of his passionate commitment to values of economic equity, social justice, frugalness, self-reliance, and environmental preservation. His work in the beautiful but economically, socially, and environmentally fragile Appalachian Mountains of southwest Virginia led to his belief that economic development can best serve society when it is based on these values and when it comes from the real needs of the people and the land. From this foundation assumption sprang Anthony's vision of ASD as an organization designed to serve as a catalyst for a healthy, equitable, socially just, environmentally renewing local economy.

From this vision emerged the working principles of ASD. These state that sustainable development: (1) is locally rooted, diversifying the economy and culture of communities and regions; (2) fits within the ecosystem, building upon natural assets, honoring limits of absorption and regeneration; (3) promotes regional self-reliance by building both individual skills and cooperative, innovative networks; (4) adds value to raw materials and shortens the distance between producers and consumers; and (5) lasts indefinitely by building the assets—ecological, human, and financial—of particular places. ASD's mission of "Developing healthy, diverse, and ecologically sound economic opportunities through education, training, and the development of cooperative networks and marketing systems" emerged from these principles, and ASD's two sustainable business units, Appalachian Harvest and Sustainable Woods, were developed to provide ASD with competitive strategic avenues to fulfill this mission. ASD's long-term goal is for each of these two sustainable strategic business units to become financially self-sufficient and to contribute 50 percent of the money necessary to cover ASD's administrative overhead.

Appalachian Harvest was the first of the two ASD sustainable strategic business units to come on line. It was founded to promote organic farming in the region and to provide an alternative income source for tobacco farmers. The success of Appalachian Harvest has been and will continue to be largely dependent on ASD's ability to recruit, educate, and support the farmers in its organic growers network. To accomplish this, ASD holds regular growers meetings and sponsors numerous seminars on organic farming. ASD also maintains a processing and packing facility where the organic produce is cleaned, graded, labeled, packed, and shipped to supermarkets in Virginia, Tennessee, and North Carolina. Appalachian Harvest sales in 2002 were $170,000—short of the $250,000 goal but 20 percent above 2001 sales. ASD currently has contracts to supply organic produce to three supermarket chains, Food City, Ukrop's Super Markets, and Whole Foods Markets. In addition to organic produce (tomatoes, peppers, cucumbers, garlic, squash, eggplant, potatoes, lettuce, and other vegetables), there are plans to include value-added products such as sauces, jams, and jellies in the Appalachian Harvest product line. There are also plans to add grass-fed, free-range, hormone-free, antibiotic-free meats, eggs, and dairy products in the future. In addition to supermarkets, ASD is working to crack the local college and university food service market.

Sustainable Woods was founded by ASD to help improve forestry management practices on private land in the region and to encourage the growth of local sustainable wood products businesses that will add value to the wood and jobs to the community. ASD has developed a three-pronged "forest-to-table" strategy for Sustainable Woods. The first prong of this strategy is to "provide outreach, education, and technical assistance for the conservation and sustainable use of private forestlands." The second is to "develop the capacity to locally process logs into kiln-dried lumber and other value-added forest products." The third is to provide for "public education and the development of markets for sustainably-produced forest products." There are three key ingredients in implementing this strategy. First, work closely with landowners and loggers to educate them and monitor them with regard to meeting ASD's sustainable wood harvesting standards. ASD currently uses its own standards for certifying sustainable wood products, but it will soon adopt the Forest Stewardship Council standards because of their broad recognitions and acceptance in the marketplace. Second, construct and operate a sawmill and kiln to process the logs into rough-cut lumber. Third, develop partnerships to supply rough-cut lumber to local builders and local producers of finished lumber, cabinets, flooring, trim, molding, wainscoting, paneling, and so forth.

ASD's success in all of its operations is dependent on how well it serves the needs of a wide variety of stakeholders. The success of ASD and the success of its networks of organic growers and landowners are clearly intertwined, and of course consumers are the ones who will ultimately determine the success of ASD, its growers, and its landowners. In order to get its products to consumers, ASD must develop networks of food retailers and finished wood products manufacturers. ASD's success is also dependent on numerous public and private granting agencies, such as the Appalachian Regional Commission, the Virginia Tobacco Commission, and the Kellogg Foundation. Employees are certainly key to ASD's success, as is the local bank that extends it a line of credit. ASD also depends on the U.S. Department of Agriculture and the Forest Stewardship Council for the standards it uses to measure the sustainability of its products, and the local community is clearly an important ASD stakeholder, as is the land the community occupies.

It should be clear from the brief description that ASD under the leadership of Anthony Flaccavento is following the path toward sustainable strategic management. Flaccavento's compelling vision based on his strong sustainability-oriented values and assumptions, the five guiding principles upon which ASD bases its strategic direction, and the strategies being pursued by the two sustainable strategic business units speak eloquently of an organization making every effort to integrate economy, society, and nature in ways that will improve the quality of life in the community for generations to come. In pursuing these strategies, ASD is serving the needs of the earth and many of its stakeholders in the business arena.

So how well is ASD doing on its road to sustainable strategic management? It has certainly had success with its recruitment, education, certification, and support strategies for growers and landowners, at least to the degree that both Appalachian Harvest and Sustainable Woods have been supplied with sufficient produce and timber to meet current demand. However, having enough growers and landowners to meet product demand will be a continuous issue to which ASD must attend, especially if it is able to grow its markets as planned. A key issue here is that recruitment of organic growers has been complicated by the new federal organic foods standards. Under the Virginia standards used over the years by ASD, land only had to be out of chemical production for one year before products grown on it could be certified organic. However, the new federal guidelines require three years out of chemical production, making it much more difficult for the farmers to reap sufficient profits from their crops. Thus, in addition to finding new markets for its organic produce, ASD is now left to find markets for an intermediate product, organically grown but not certified produce. Unlike Appalachian Harvest with its potential supply problems, finding landowners to supply the timber for Sustainable Woods may not be much of a problem in the future. The Nature Conservancy, one of ASD's longtime partners, has just added 5,750 forested acres to its Conservation Forestry Program—a sustainable forest management program—just a few miles from the Sustainable Woods sawmill and kiln. This could provide a virtually endless supply of logs for Sustainable Woods.

With regard to its processes for creating, transforming, and distributing its products, ASD also has the capacity to meet current demand. The capacity of the Appalachian Harvest packing facility was stretched to its limits in 2002 (its first year of operation), but its capacity has been doubled for 2003. A larger issue for Appalachian Harvest in the future may be transportation of its produce to supermarkets. Currently Appalachian Harvest makes its own deliveries (many of which are four to six hours from the packing facility). However, growth will likely stretch its current transportation capacity, and ASD will face a decision regarding whether to expand its truck fleet or to contract for transportation services. The sawmill and kiln are also sufficient for current demand. In fact, storage capacity for dried ready-to-sell lumber is currently stretched to its limit. The kiln is solar powered with a backup scrapwood burning boiler. The original intent was for the kiln to be primarily solar, but some problems have arisen in connection with this. The first problem is typical for solar technology—there is not enough consistent sunlight, especially during the winter months. Second, the original covering on the solar panels did not retain enough heat and was prone to tearing. This material has recently been replaced with more durable heat retaining material, so it is hoped that this will be a less significant problem in the future. Third, the timing of the solar drying process is not as predictable as more traditional drying methods. Because of these problems, Sustainable Woods relies a great deal on its backup boiler. The boiler uses scrap wood from the sawing process, so in that sense this practice is sustainable. However, the boiler does release some CO_2 into the atmosphere.

Thus, ASD has both an adequate supply of inputs into its two sustainable strategic business units to meet current demand and an adequate capacity to process those inputs into market-ready outputs. However, ASD's most critical issue is embedded in the phrase, "current demand." ASD has been continuously plagued with finding enough buyers for its organic produce and rough-cut lumber. It has had some success in this regard. As mentioned above, ASD sold $170,000 worth of organic produce in 2002 to three supermarket chains. However, sales of produce do not currently cover all of Appalachian Harvest's employee, processing, and transportation costs. Sustainable Woods is also not covering its costs. Further, the market for sustainable lumber has been slower to develop than the organic produce market. ASD believes there are four reasons for this slow development: the market for sustainable wood is not as mature as the market for organic food, inventory is limited in both type and quality of wood, the price of sustainable lumber is higher than for other lumber, and many finished goods manufacturers have already established buying patterns with other suppliers. In order to improve its market position, Sustainable Woods needs more brand recognition and more exposure for its products. To accomplish this, ASD is currently seeking high visibility projects—such as churches, libraries, and schools—where products made from its wood are on public display. One such project, the Jubilee House (a retreat attached to a local Catholic church), displays beautiful flooring, trim, exposed beams, and wainscoting made from Sustainable Woods lumber, and ASD would like to have many more like this.

Thus, in a nutshell ASD has serious problems at the marketing end of the value chain. The marketing efforts at ASD have been genuine but inconsistent, and they have been plagued with responsibility-designation and turnover problems. ASD has sought and received marketing help from a local university, and it has employed someone responsible for marketing organic produce for the past couple of years. With regard to Sustainable Woods, the forester responsible for recruiting landowners and monitoring timber harvesting standards is also responsible for marketing the lumber—giving her responsibility at both ends of the value chain. There is a need for an employee whose primary job is lumber sales, as well as someone to market the Appalachian Harvest produce to stores, institutions, and so forth, and someone else to recruit farmers for the Appalachian Harvest growers network. Nonetheless, to this point in its history, ASD's marketing efforts have been inadequate, with the result that ASD is not achieving its primary economic goal—for each sustainable strategic business unit to be financially independent and to be net contributors to ASD's administrative overhead. As a result, ASD has had continuous economic problems, including recently having to cut all employees' hours by 30 percent, effective until further grant funds are available. With regard to this heavy reliance on grant funding to pay its employees and cover its costs, Anthony Flaccavento has proved to be a master over the years at securing both restricted and unrestricted grants that have allowed ASD to remain afloat and do its important work, and grants will (and should) always be an important source of funds for ASD because of what it is and what it does. However, to this point, for all of its good works and good intentions with regard to revitalizing the greater social system and protecting the ecosystem, ASD has not, according to its own definition, achieved economic sustainability.

Bill Ford and Ford Motor Company

Recall from Chapter 1 that Bill Ford, chairman and CEO of Ford Motor Company, has been referred to as a rebel, a radical, too green, and too idealistic.[2] Many have been skeptical about his ability to lead Ford Motor Company in the hard-knocks world of competitive global business. Clearly very smart, he held several key jobs in the company for twenty-three years prior to becoming chairman and CEO, which prepared him for his appointment. However, many are still nervous, especially when he touts his sustainability agenda. They are simply not very comfortable when he questions the future of personal vehicle transportation or ponders the ethical implications of his firm's participation in the sport utility vehicle (SUV) market. He has nonetheless been relentless in his efforts to lead Ford toward sustainable strategic management. As he noted in Ford's *2001 Corporate Citizenship Report*, "We want to continue to provide the world with mobility by making it affordable in every sense of the word—economically, environmentally, and socially."

Ford has identified five corporate level sustainable strategic management priorities. The first priority is to act on a strong commitment to human rights. In carrying out this strategic initiative, Ford has established a human rights strategy development team made up of representatives from the departments of Human Resources, Marketing and Sales, Purchasing, Manufacturing, Emerging Markets, and the Office of General Counsel. This team has instituted a dialogue with a wide variety of external human rights groups to develop a human rights policy for the firm. The policy addresses child, prison and forced labor, health and safety, compensation, freedom of association, harassment, and discrimination issues.

Ford's second corporate sustainable strategic priority is to focus on sustainable mobility. In pursing this priority, Ford is partnering with eleven other firms in the automobile and energy industries to form the core group of companies participating in the World Business Council on Sustainable Development's (WBCSD) Sustainable Mobility Project. These companies are conducting a dialogue with stakeholders from all parts of the globe and all levels of society to achieve worldwide sustainable mobility by 2030. This will require creating mobility systems that provide free movement of people and goods without contributing to pollution, resource depletion, traffic congestion, and social inequity. The first phase of the project was to analyze current mobility patterns in both the developed and developing worlds. Thirteen sustainable mobility criteria were measured as part of this analysis: access to means of mobility, equity of access, mobility infrastructure, cost of freight, congestion, nonrenewable energy use, greenhouse gas emissions, noise, safety, community disruption, waste, conventional emissions, and other environmental impacts. Of these thirteen criteria, only two were rated acceptable in the developed world—access to means of mobility and cost of freight. The remainder were rated either needs improvement or unacceptable. In the developing world the news was even worse. None of the criteria was rated acceptable, and eight of the thirteen were rated unacceptable. These data are being used for the second phase of the project, which is to determine the "grand challenges" regarding creating sustainable transportation systems in the future. Phase three is the development of a final report that includes recommendations for meeting these challenges in order to make sustainable mobility a reality by 2030.

The third sustainable strategic priority at Ford is to develop a comprehensive climate change strategy. To accomplish this, Ford has established a three-pronged strategy that develops a climate change inventory and baseline, considers a wide range of measures to reduce greenhouse emissions, and from these measures establishes a long-term climate change strategy. In developing this strategy, Ford has entered several strategic partnerships, including the sustainable mobility partnership discussed above and a partnership with BP Amoco to supply low-sulfur gasoline for Ford to use to fill its new cars at the factory. Ford is also working to develop more accurate and more sophisticated environmental modeling capabilities through Ford Research Laboratory, and it is working in partnership with several other firms to develop fuel cell technology (discussed in more detail below).

The fourth sustainable strategic priority for Ford is to establish a set of business principles designed to provide an underlying value system upon which to build the firm. Rather than crafting these principles at the top management level and autocratically distributing them to employees, Bill Ford and his team of strategic managers have established an open dialogue process designed to involve organizational members from all levels and all strategic business units across the globe in the business principle development process. They felt that such a process would allow them to explore the business principles from all perspectives before finalizing them, and help to ensure that the principles would be ingrained in Ford's organizational culture from their inception. In the process, a draft of six broad principles related to financial health, products and customers, community, environment, accountability, and quality relationships was given to over 4,000 Ford employees for feedback and suggestions. These employees were very supportive of the content and spirit of the six principles in the draft proposal, but they were concerned about how the principles would be made operational. Based on this feedback, Ford's strategic management team decided that further analysis and discussion were desirable. Thus, the principles were not finalized until 2002, two years after they were proposed. During that time, open discussion on the principles was encouraged and a broad consensus was achieved.

The fifth sustainable strategic priority for Ford is to seek new solutions through technology and partnerships. Ford's strategic management team believes that this priority is critical if the firm is to meet its first three strategic priorities, human rights, sustainable mobility, and climate change. Ford is pouring huge amounts of resources into sustainable technologies designed to improve the economic, social, and environmental performance of its products and manufacturing processes, and many of these new technologies are being developed via partnerships with other auto manufacturers, energy companies, interest groups, governments, and so forth.

With regard to its products, Ford is currently developing several vehicles that use alternative fuels, including vehicles powered by hydrogen fuel cells, hybrid electric power, propane, natural gas, and ethanol. With regard to fuel cell development, Ford is a part of two research partnerships—the California Fuel Cell Partnership and the U.S. government's Project FreedomCAR. Another ambitious product development strategy at Ford is the development of the Model U concept car. The Model U is a vehicle of the future made from technical and biological nutrients (see Chapter 5) that can either be recycled and reused indefinitely or returned to the soil. In this regard, Ford is currently working in partnership with two suppliers. One is Milliken & Company, with whom Ford is working to develop a technical nutrient called ecoeffective polyester to be used in the Model U's seats, steering wheels, headrests, door trim, and armrests. Another supplier Ford is partnering with in this project is Cargill Dow; together these firms are developing a corn-based biological nutrient (mentioned in Chapter 9) called polylactide (PLA) for use in the Model U's canvas roof and carpeting.

With regard to manufacturing, Ford's most ambitious and visible sustainability project is the $2 billion renovation of its famous Rouge plant in Dearborn, Michigan. Ford hired William McDonough and his associates to design the Rouge as an ecoeffective manufacturing facility. When completed, the plant will have a 500,000-square-foot living roof, a system of natural plants on the grounds surrounding the plant to rid the soil of contaminants, a system of swales (shallow ditches seeded with indigenous plants) and porous paving designed to manage storm water runoff, trellises of flowering plants to shade and cool the assembly and office areas, solar and fuel cells to provide alternative energy sources for the plant, and a system of trees and plants around the plant designed to attract birds and other species to the Rouge's grounds. The plant is also designed to provide a healthy, productive, supportive work environment. The temperature is comfortable year round, and the plant has natural light, safety overhead walkways, team rooms, and natural areas in which people can congregate. In addition to the Rouge, Ford has sustainable production plant projects going on at its Windsor Engine Plant in Windsor, Ontario, and the Complexo Industrial Ford Nordeste in Camacari, Brazil. The Rouge and other sustainable manufacturing projects are designed to serve as models for Ford's long-term goal of having nothing but sustainable manufacturing facilities.

In pursuing its five corporate level sustainable strategic priorities, Ford focuses significant attention on creating sustainable relationships with a wide array of stakeholders, including employees, customers, investors, automobile dealers, suppliers, communities, and representatives of civil society (interest groups that represent society and nature). With regard to employees, Ford is committed to a safe work environment that is economically, socially, environmentally, and spiritually satisfying. With regard to customers, Ford is committed to providing safe, high quality, innovative products at a fair price, and to providing accurate information to its customers while protecting their privacy. Ford is working to increase the number of minority owners in its dealer network, and it is working in partnership with several companies in Sweden to develop environmentally sustainable automobile dealerships (known as GreenZone dealerships). Ford recognizes that its suppliers are key to its ability to achieve sustainability, so it is working in partnerships with many suppliers (such as Milliken, Cargill Dow, and BP mentioned above) to improve their environmental and social performance. Ford is committed to providing reasonable financial returns to its investors via sustainable strategic management strategies that provide for growth within the context of Ford's business principles. It is also committed to providing economic opportunity, safety, a high quality of life, and a clean environment in the 110 communities in twenty-five countries on six continents where it operates, and it is committed to entering into dialogue with organizations representing civil society (Greenpeace, Sierra Club, the World Business Council for Sustainable Development, Lawyers Committee for Human Rights,

INFORM, and so forth) in order to better understand the social and environmental risks and opportunities facing the firm.

Ford has had and will continue to have many successes in its sustainable strategic management efforts. For example, it was ranked as one of the top fifty companies for minorities by *Fortune* magazine in 2002, and *DiversityInc.com* ranked it number one for diversity in 2003. Nonetheless, despite its ambitious sustainability vision, its values, its business principles, its SSM strategies, its successes, and its commitments to stakeholders, Ford Motor Company is facing some very serious economic issues. In 2001 Ford experienced its first quarterly loss in ten years, it recalled 13 million Firestone tires that had been original equipment on Ford Explorers, and it reduced its common stock dividend. In 2002 its stock hit a ten-year low, it again lowered dividends on common stock, its pension fund experienced a $6.5 billion shortfall, its debt reached $162 billion, and it had to restructure to the tune of 35,000 lost jobs in an effort to improve efficiency, productivity, and the economic bottom line. With economic troubles like these looming over the company for the foreseeable future, many wonder if the sustainability focus at Ford is realistic. Bill Ford addressed this point in the *2001 Corporate Citizenship Report*, saying, "A legitimate question to ask is whether our intense focus on the economic side of our business will distract us from our environmental and social efforts. . . . Difficult business conditions make it harder to achieve the goals that we set for ourselves in many areas, including corporate citizenship. But that doesn't mean we will abandon our goals or change our direction."

The Real Triple Bottom Line

So, here we have Anthony Flaccavento, strategic leader of a small not-for-profit organization, and Bill Ford, strategic leader of one of the world's largest corporations, floating in essentially the same boat. Both are clearly committed to visions of sustainable strategic management, and both face significant economic obstacles in fulfilling their visions. Further, the situations faced by Flaccavento and Ford are indicative of the situations faced by most strategic managers in organizations pursuing sustainable strategic management in today's highly competitive business environment. For all of the depth they have in what they believe and value, for all of the passion they invest in their visions, for all of their good intentions and effective leadership, and for all of their ability to craft and implement effective SSM strategies, Anthony Flaccavento and Bill Ford, and other strategic managers seeking sustainable strategic management, will, for the foreseeable future, face one dominant dimension—economic success, the bear that rules the triple bottom line. The fact is, in today's world, firms can pay low wages, do little for their communities, contribute wasteful consumer products to society, and minimally meet environmental regulations, but if they cover their costs and earn occasional profits, they can survive. Unfortunately, the opposite is not true. Firms can pay high wages and encourage employee participation, growth, and development, be partners in making their communities and their world better places in which to live, produce and sell socially and environmentally responsible products and services, and consistently take a proactive approach to preventing environmental harm, but if they do not cover their costs and generate enough financial gain to keep their firms competitive, their survival will be threatened. Thus, in today's business world it is good to be concerned about social responsibility and ecological health, but the ability to integrate them into the strategic fabric of organizations is limited by the impact these have on the economic bottom line of the firm.

Coming Full Circle

From the outset of the book we have contended that successfully implementing sustainable strategic management will be difficult without a significant paradigm shift—a shift from the belief that the economy operates in a closed circular flow that is separate from and independent of the greater social system and ecosystem, to the belief that the economy is an open living system that is a part of and interconnected to the greater social system and ecosystem. In the interim we have built what we believe is a solid foundation that allows us to understand, formulate, implement, and evaluate sustainable strategic management processes. Let us review the elements of this foundation.

We began the book with short introductions of two committed sustainable strategic managers, Anthony Flaccavento and Bill Ford. Using their stories as springboards, we built the case that sustainable strategic management is not just a new way of doing business; it is a new way of thinking about business that will require widespread transformational change in business organizations. We then provided an in-depth look at the concept of sustainability, presenting some of its underlying scientific principles, discussing its economic, social, and ecological dimensions, presenting some of its ethical foundations, and examining some of the tenets of a sustainability worldview. Next, we defined sustainable strategic management, relating it to the popular triple-bottom-line concept, and we discussed the roles and responsibilities of CEOs and boards of directors in leading organizations toward sustainable strategic management.

We followed this with a discussion of external environmental analysis as it relates to identifying sustainability-based threats and opportunities; we focused on analyzing the interactions among population growth, economic growth, and technological efficiency in order to understand the macrosustainability issues facing organizations, and we focused on coevolutionary industry analysis and multiple scenario analysis as useful means of analyzing the sustainability dimensions of the competitive (industry) environment. Next, we discussed how, when modified to account for the greater social system and ecosystem, both value-chain analysis and stakeholder analysis can be useful tools in determining organizational strategic advantages. We then provided an in-depth examination of the content of SSM strategies at the functional, competitive, and corporate levels, concluding that successful sustainable strategic management will require a corporate-level portfolio of SSM strategies that allows firms to focus on "what can be" economically, socially, and ecologically. Next, we provided an in-depth look at the controlled, value-driven cognitive processes required for complex strategic choices, and we presented a set of instrumental values that we believe can support sustainable strategic management decisions in organizations. We then presented a model of the sustainable strategic management process, highlighting how organizations with sustainability-centered value systems that pay attention to social and environmental issues and recognize the legitimacy of the earth's stakeholders can be said to "stand for sustainability."

We followed this with a discussion of several sustainable strategic management systems that we think are critical for effective implementation of SSM strategies, including sustainable management systems, research and development, procurement, operations, infrastructures marketing, and so forth. We discussed how effective implementation of SSM strategies requires sustainability-based organizational cultures, human resource systems, learning structures, and disruptive technologies, and we presented some important systems that are necessary to evaluate the effectiveness of SSM strategies, including performance measurement systems, finance and accounting systems, information systems, and reporting systems.

Now, here in the last chapter, we have revisited and expanded the stories of ASD and Ford Motor Company under the strategic leadership of Anthony Flaccavento and Bill Ford. We did this because we believe that both Flaccavento and Ford represent excellent examples of strategic managers leading their firms toward sustainable strategic management as described in this book. Both demonstrate SSM intelligence in their deeply rooted and strongly held sustainability-centered values and visions. Both ensure that their organizations purposefully serve the social and ecological as well as the economic needs of their stakeholders. Both are committed to eco- and socioeffectiveness, wishing to leave the land and its people better off in the future for their efforts. Both ASD and Ford are consciously working toward operating within a type III closed-loop value chain, pursuing sustainable designs, energy sources, materials, production processes, products, services, and so forth, often in partnership with stakeholders from the economic, social, and ecological realms. Both have developed a corporate portfolio of SSM strategies that focuses on "what can be." In short, there is little doubt that Anthony Flaccavento, Bill Ford, and their organizations "stand for sustainability." However, none of this has made these strategic managers and their organizations any less immune to the mighty grip of the economic bear, and in this they are certainly not alone. We chose ASD and Ford from among dozens of organizations that are currently pursuing sustainable strategic management. All of them want to contribute to the greater economic, social, and ecological good, and all of them are most constrained in such pursuits by economic concerns.

So we have come full circle. We have demonstrated via ASD, Ford, and others that regardless of the values, intentions, and actions of SSM-intelligent strategic managers, successfully implementing sustainable strategic management may never be fully realized until the economic paradigm shifts to a more open living system framework that reflects a more egalitarian interconnection among the planet's economic system, social system, and ecosystem. This will require rethinking and reshaping the current economic story that stresses centrality, dominance, control, instrumentality, and immediacy with a new economic story that stresses interconnectedness, reciprocity, collaboration, coevolution, spirituality, and posterity.[3] As daunting as this prospect seems, we are optimistic that such a shift can occur and, in fact, is occurring. Our experience in this field over the past two decades or so tells us that Anthony Flaccavento and Bill Ford are part of a growing cadre of strategic managers who are showing a willingness, even during economic hard times, to confront the economic bear face to face and say, "Sure, we are going to feed you, but we are going to learn how to do it on our own sustainable terms." With growth in the number of strategic managers who think this way, we believe that the paradigm will continue to shift toward the open living system economy, increasing the possibility that a sustainable world awaits future generations.

Notes

Notes to Chapter 1

1. This quote has appeared in several articles. It may be found in B. Morris, N. Watson, and P. Neerings, "Idealist on Board," *Fortune*, April 3, 2000; available at www.fortune.com/articles/0,15114,371774,00.html.

2. See T. Kiuchi and B. Shireman, *What We Learned in the Rainforest* (San Francisco: Barrett-Koehler, 2002) for a detailed discussion of the lessons that businesses can learn from nature.

3. For an in-depth discussion of the stages and processes of human and social development, please see D. Beck and C. Cowan, *Spiral Dynamics: Mastering Values, Leadership, and Change* (Cambridge, MA: Blackwell, 1995); and K. Wilber, *A Theory of Everything: An Integral Vision for Business, Politics, Science, and Spirituality* (Boston, MA: Shambhala, 2000).

4. For a more in-depth examination of the organizational culture, organizational change, the degrees of change, and transformational change, please see the following: J. Bartunek and M. Moch, "First-Order, Second-Order, and Third-Order Change and Organization Development Interventions: A Cognitive Approach," *Journal of Applied Behavioral Science* 23 (1987): 483–500; R. Beckhard and W. Pritchard, *Changing the Essence* (San Francisco: Jossey-Bass, 1992); and E. Schein, *Organizational Culture and Leadership* (San Francisco: Jossey-Bass, 1985).

5. Some of the literature on the role of transformational change in sustainable strategic management can be found in the following: D. Dunphy, A. Griffiths, and S. Benn, *Organizational Change for Corporate Sustainability* (London: Routledge, 2002); R. Freeman, J. Pierce, and R. Dodd, *Environmentalism and the New Logic of Business* (Oxford: Oxford University Press, 2000); T. Gladwin, J. Kennelly, and T. Krause, "Shifting Paradigms for Sustainable Development: Implications for Management Theory and Research," *Academy of Management Review* 20, no. 4 (1995): 874–907; J. Post and B. Altman, "Managing the Environmental Change Process: Barriers and Opportunities," *Journal of Organizational Change Management* 7, no. 4 (1994): 64–81; J. Post and B. Altman, "Models for Corporate Greening: How Corporate Social Policy and Organizational Learning Inform Leading-Edge Environmental Management," in *Research in Corporate Social Policy and Performance*, ed. J. Post, 3–29 (Greenwich, CT: JAI Press, 1992); W. Stead and J. Stead, "Can Humankind Change

the Economic Myth? Paradigm Shifts Necessary for Ecologically Sustainable Business," *Journal of Organizational Change Management* 7, no. 4 (1994): 15–31; and G. Throop, M. Starik, and G. Rands, "Sustainable Strategy in a Greening World: Integrating the Natural Environment into Strategic Management," in *Advances in Strategic Management*, ed. P. Shrivastava, A. Huff, and J. Dutton, 63–92 (Greenwich, CT: JAI Press, 1993).

6. See Post and Altman, "Models for Corporate Greening," 13.

Notes to Chapter 2

1. This well-known definition was originally put forth by the World Commission on Environment and Development in *Our Common Future* (Oxford: Oxford University Press, 1987), 8.

2. The well-established tenets of systems theory can be found in a wide variety of sources. The discussion of living systems in this chapter was developed from the following sources: K. Boulding, "General Systems Theory: The Skeleton of Science," *Management Science* 2, no. 3 (1956): 197–208; E.F. Schumacher, *A Guide for the Perplexed* (New York: Harper and Row, 1977); R. Sheldrake, *The Greening of Science and God* (New York: Bantam Books, 1991); J. Van Gigch, *Applied General Systems Theory* (New York: Harper and Row, 1978); and K. Wilber, *Quantum Questions* (Boston: New Science Library, 1985).

3. This discussion was developed primarily from the following sources: J. Lovelock, *Gaia: A New Look at Life on Earth* (London: Oxford University Press, 1979); J. Lovelock, *The Ages of Gaia* (New York: Bantam Books, 1988); L. Margulis and G. Hinkle, "The Biota and Gaia: 150 Years of Support for Environmental Sciences," in *Scientists on Gaia*, ed. S. Schneider and P. Boston, 11–18 (Cambridge: MIT Press, 1991); P. Ehrlich, "Coevolution and Its Applicability to the Gaia Hypothesis," in *Scientists on Gaia*, 19–22; and L.E. Joseph, *Gaia: The Growth of an Idea* (New York: St. Martin's Press, 1990).

4. The sources from which this discussion was developed provide thorough analyses of the relationship between entropy and economics. These include: N. Georgescu-Roegen, *The Entropy Law and the Economic Process* (Cambridge: Harvard University Press, 1971); and F. Capra, *The Turning Point* (New York: Bantam Books, 1982).

5. This discussion was primarily derived from: R. Ayres, "Industrial Metabolism," in *Technology and Environment*, ed. J. Ausubel and H. Sladovich, 23–49 (Washington, DC: National Academy Press, 1989); and R. Ayres, "Industrial Metabolism: Theory and Policy," in *The Greening of Industrial Ecosystems*, ed. B. Allenby and D. Richards, 23–37 (Washington, DC: National Academy Press, 1994).

6. The data in this section (unless otherwise cited) are from the Worldwatch Institute's *Signposts 2002*, a CD-ROM containing four of the institute's major publications, *State of the World 2001 and 2002*, and *Vital Signs 2001 and 2002*, and a plethora of other articles and databases on a wide variety of topics related to sustainability. This is an outstanding source for anyone interested in delving more deeply into the root issues related to sustainability.

7. Information regarding the digital divide is from the Markle Foundation, *Creating a Development Dynamic: Final Report of the Digital Opportunity Initiative*, July 2001; available at www.markle.org/news/DOIReportWord.doc.

8. See M. Shuman, *Going Local* (New York: Free Press, 1998).

9. For a thorough discussion of these stages of industrial development, see D. Richards, B. Allenby, and R. Frosch, "The Greening of Industrial Ecosystems: Overview and Perspective," in *The Greening of Industrial Ecosystems*, 1–19.

10. This is from the following source, which provides an excellent discussion of what a myth is and how it relates to modern society: J. Campbell (with B. Moyers), *The Power of Myth* (New York: Doubleday, 1988).

11. The ideas summarized here are from some prominent contributors to this way of thinking, and an in-depth discussion of their ideas can be found in the following sources: T. Berry, *The Dream of the Earth* (San Francisco: Sierra Club Books, 1988); Capra, *The Turning Point*; R. Carson, *Silent Spring* (Boston: Houghton Mifflin, 1962); H. Daly, *Steady State Economics* (San Francisco: W.H. Freeman, 1977); H. Daly and J. Cobb, *For the Common Good* (Boston: Beacon Press, 1989); A. Leopold, *A Sand County Almanac* (London: Oxford University Press, 1949); L. Milbrath, *Envisioning a Sustainable Society* (Albany: State University of New York Press, 1989); Schumacher, *A Guide for the Perplexed*; and E.F. Schumacher, *Small Is Beautiful: Economics as if People Mattered* (New York: Harper and Row, 1973).

12. The discussion of the ideas of Aldo Leopold comes from his best-known work: *A Sand County Almanac*.

13. Leopold, *A Sand County Almanac*, viii–ix.

14. Ibid., 203.

15. Ibid., 221.

16. The information on Schumacher's life is from a biography written by his daughter: B. Wood, *E.F. Schumacher: His Life and Thought* (New York: Harper and Row, 1984). The discussion of his ideas comes primarily from two sources: Schumacher, *A Guide for the Perplexed*; and Schumacher, *Small Is Beautiful*.

17. The summary of the economic-wealth and sustainability mental models presented here was developed by T. Gladwin, J. Kennelly, and T. Krause, "Shifting Paradigms for Sustainable Development," *Academy of Management Review* 20, no. 4 (1995): 874–907.

Notes to Chapter 3

1. Sources for this paragraph include: S. Hart, "A Natural Resource-Based View of the Firm," *Academy of Management Review* 20, no. 4 (1995): 966–1014; W. McDonough and M. Braungart, *Cradle to Cradle* (New York: North Point Press, 2002); P. Shrivastava, "The Role of Corporations in Achieving Ecological Sustainability," *Academy of Management Review* 20, no. 4 (1995): 936–60; M. Starik and G. Rands, "Weaving an Integrated Web: Multilevel and Multisystem Perspectives of Ecologically Sustainable Organizations," *Academy of Management Review* 20, no. 4 (1995): 908–35; W. Stead and J. Stead, "An Empirical Investigation of Sustainability Strategy Implementation in Industrial Organizations," in *Research in Corporate Performance and Policy*, Supplement 1, ed. J. Post, vol. ed., D. Collins and M. Starik, 43–66 (Greenwich CT: JAI, 1995); W. Stead and J. Stead, *Management for a Small Planet: Strategic Decision Making and the Environment* (Newberry Park, CA: Sage, 1992); and G. Throop, M. Starik, and G. Rands, "Sustainable Strategy in a Greening World: Integrating the Natural Environment into Strategic Management," in *Advances in Strategic Management*, vol. 9, 63–92 (Greenwich, CT: JAI, 1993). Together these references provide an excellent perspective on the development of the concept of sustainable strategic management.

2. The term "triple bottom line" was introduced by J. Elkington, *Cannibals with Forks* (Oxford, UK: Capstone, 1997).

3. See J. Schmidt, "Corporate Excellence in the New Millennium," *Journal of Business Strategy* (November/December 1999): 39–43.

4. A few of the most influential books with regard to CEO leadership roles and responsibilities include: J. Collins, *Good to Great* (New York: HarperBusiness, 2001); J. Collins and J. Porras, *Built to Last: Successful Habits of Visionary Companies* (New York: HarperBusiness, 1994); J. Moore, *The Death of Competition: Leadership and Strategy in the Age of Business Ecosystems* (New York: HarperBusiness, 1996); T. Peters and R. Waterman, Jr., *In Search of Excellence* (New York: Harper and Row, 1982); P. Senge, *The Fifth Discipline: The Art and Practice of the Learning Organization* (New York: Doubleday Currency, 1990). Information from these sources is woven throughout this section on leadership.

5. This is a central point in William McDonough's philosophy of ecoeffectiveness and a point he extols constantly in his writings and teachings. For more information on ecoeffectiveness, see McDonough and Braungart, *Cradle to Cradle*.

6. See Collins, *Good to Great*; Collins and Porras, *Built to Last*; Senge, *The Fifth Discipline*.

7. Our information on enterprise strategy was compiled from the following sources: I. Ansoff, "The Changing Shape of the Strategic Problem," in *Strategic Management*, ed. D. Schendel and C. Hofer, 30–44 (Boston: Little Brown, 1979); E. Freeman, *Strategic Management: A Stakeholder Approach* (Boston: Pitman, 1984) (the purpose of enterprise strategy as quoted in this paragraph can be found on page 90 of this seminal work); E. Freeman and D. Gilbert, Jr., *Corporate Strategy and the Search for Ethics* (Englewood Cliffs, NJ: Prentice-Hall, 1988); L. Hosmer, "Strategic Planning as if Ethics Mattered," *Strategic Management Journal* 15 (1994): 17–34.

8. See J. Stead and E. Stead, "Eco-Enterprise Strategy: Standing for Sustainability," *Journal of Business Ethics* 24, no. 4 (2000): 313–29.

9. The discussion of stakeholder theory and research in this and the following paragraph was compiled from the following sources: B. Agle, R. Mitchell, and J. Sonnenfeld, "Who Matters to CEOs? An Investigation of Stakeholder Attributes and Salience, Corporate Performance, and CEO Values," *Academy of Management Journal* 42, no. 5 (1999): 507–25; S. Berman, A. Wicks, S. Kotha, and T. Jones, "Does Stakeholder Orientation Matter? The Relationship Between Stakeholder Management Models and Firm Financial Performance," *Academy of Management Journal* 42, no. 5 (1999): 488–506; A. Carroll, "Stakeholder Thinking in Three Models of Management Morality: A Perspective with Strategic Implications," in *Understanding Stakeholder Thinking*, ed. J. Nasi, 47–74 (Helsinki: LSR, 1995) ; M. Clarkson, "A Stakeholder Framework for Analyzing and Evaluating Corporate Social Performance," *Academy of Management Review* 20, no. 1 (1995): 92–117; T. Donaldson and L. Preston, "The Stakeholder Theory of the Corporation: Concepts, Evidence, and Implications," *Academy of Management Review* 20, no. 1 (1995): 65–91; Freeman, *Strategic Management*; Freeman and Gilbert, *Corporate Strategy and the Search for Ethics*; Hosmer, "Strategic Planning as if Ethics Mattered"; T. Jones, "Instrumental Stakeholder Theory: A Synthesis of Ethics and Economics," *Academy of Management Review* 20, no. 2 (1995): 404–37; S. Ogden and R. Watson, "Corporate Performance and Stakeholder and Customer Interests in the U.K. Privatized Water Industry," *Academy of Management Journal* 42, no. 5 (1999): 526–38.

10. See D. Weidman, "Redefining Leadership for the Twenty-First Century," *Journal of Business Strategy* (September–October 2002): 16–18.

11. See T. Hayes, "The New Bottom Line for Business," *Journal of Business Strategy* (November–December 2002): 34–36; and Weidman, "Redefining Leadership for the Twenty-First Century."

12. These characteristics of learning organizations are derived from Senge, *The Fifth Discipline*.

13. See D. Goleman, *Emotional Intelligence* (New York: Bantam Books, 1996) for a thorough discussion of emotional intelligence.

14. The information for this section is largely drawn from D. Zohar and I. Marshall, *Spiritual Intelligence: The Ultimate Intelligence* (New York: Bloomsbury, 2000).

15. See Elkington, *Cannibals with Forks*.

16. See P. Luoma and J. Goodstein, "Stakeholders and Corporate Boards: Institutional Influence on Board Composition and Structure," *Academy of Management Journal* 42, no. 5 (1999): 553–63.

17. For a thorough discussion of the issue of interlocking directors, see "Web of Board Members Ties Together Corporate America," *USA Today*, November 25, 2002: B1–B3.

18. Many contend that the recent corporate scandals are triggering a revolution in corporate governance. For more information, see D. Finegold, E. Lawler, and J. Conger, "Building a Better Board," *Journal of Business Strategy* (November/December 2001): 33–37; Hayes, "The New Bottom Line for Business"; and L. Lavelle, "The Best and Worst Boards," *Business Week*, October 7, 2002: 104–14.

19. See B. Garratt, *The Fish Rots from the Head: The Crisis in Our Boardrooms—Developing the Crucial Skills of the Competent Director* (New York: HarperBusiness, 1996).

20. See A. Kleiner, G. Roth, and N. Kruschwitz, "Should a Company Have a Noble Purpose?" *Across the Board* (January 2001): 18–26.

21. The examples in this section are from AOL Time Warner's 2002 Social Responsibility Report; B. Smidt, "Sustained Growth for Sustainable Growth," *Chemical Week* 162, no. 44 (2000): 33–34; and B. Williams, "Comment: Is Shell Report 2000's Sustainable Development Focus an Anomaly or a Sign of the Future?" *Oil and Gas Journal* 98, no. 46 (2000): 74–80.

22. See Nike's 2001 *Corporate Responsibility Report*.

Notes to Chapter 4

1. E.F. Schumacher, *Good Work* (New York: Harper and Row, 1979): 5. This book contains a collection of essays and presentations Schumacher gave in North America shortly before his death.

2. I. Ansoff, "The Changing Shape of the Strategic Problem," in *Strategic Management*, ed. D. Schendel and C. Hofer, 30–44 (Boston: Little Brown, 1979).

3. These data and the remainder of the data presented in this section are from *Tomorrow's Markets* (Washington, DC: World Resources Institute, United Nations Environment Programme, World Business Council for Sustainable Development, 2002).

4. It should be noted here that F. David, *Strategic Management* (Upper Saddle River, NJ: Prentice-Hall, 2003) and J. Pearce and R. Robinson, *Strategic Manage-*

ment (Boston: McGraw-Hill-Irwin, 2003) both include the ecological sector in macroenvironmental analysis.

5. The effects of population growth have been the subject of many outstanding books and articles. The information regarding population in this chapter is primarily from the following: Worldwatch Institute, *Signposts 2002* (a CD-ROM containing four of their major publications, *State of the World 2001 and 2002*, and *Vital Signs 2001 and 2002*); P. Ehrlich and A. Ehrlich, *The Population Explosion* (New York: Simon and Schuster, 1990); and H. Kendall and D. Pimentel, "Constraints on the Expansion of the Global Food Supply," *Ambio* 23, no. 3 (1994): 198–205.

6. The information regarding technology is largely from: Worldwatch Institute, *Signposts 2002*; S. El Serafy, "The Environment as Capital," in *Ecological Economics*, ed. R. Costanza, 168–75 (New York: Columbia University Press, 1991); R. Goodland, H. Daly, and S. El Serafy, *Population, Technology, and Lifestyle* (Washington, DC: Island Press, 1992).

7. The data in this section are from Worldwatch Institute, *Signposts 2002*.

8. The data in this section are from Worldwatch Institute, *Signposts 2002*. See also A. Hoffman, "An Uneasy Rebirth at Love Canal," *Environment* 37, no. 2 (1995): 3–9, 25–31.

9. Our arguments in this section are largely from M. Porter, *The Competitive Advantage of Nations* (New York: Free Press, 1990). We also drew upon other writings of Porter in this section, including: M. Porter, "America's Green Strategy," *Scientific American* (April 1991): 168; and M. Porter and C. van der Linde, "Green and Competitive: Ending the Stalemate," *Harvard Business Review* 73, no. 5 (1995): 120–34.

10. Our arguments in this section are taken largely from the following works of Herman Daly (and his colleague John Cobb): H. Daly, *Steady-State Economics*, 2d ed. (Washington, DC: Island Press, 1991); H. Daly, "The Perils of Free Trade," *Scientific American* (November 1993): 50–57; and H. Daly and J. Cobb, Jr., *For the Common Good*, 2d ed. (Boston: Beacon Press, 1994).

11. Our arguments in this section are taken largely from M. Shuman, *Going Local* (New York: Free Press, 1998).

12. The case of Johnson County, Tennessee, is covered in depth in E. Stead, J. Stead, and D. Shemwell, "Community Sustainability in the Southern Appalachian Region of the USA: The Case of Johnson County, Tennessee," in *Research in Corporate Sustainability*, ed. S. Sharma and M. Starik, 61–84 (Northhampton, MA: Edward Elgar, 2003).

13. See M. Porter, "How Competitive Forces Shape Strategy," *Harvard Business Review* (April–May 1979): 137–45.

14. Our arguments in this section are largely taken from J. Moore, *The Death of Competition* (New York: HarperBusiness, 1996).

15. J. Mahon and R. McGowan, "Modeling Industry Political Dynamics," *Business and Society* 37, no. 4 (1998): 391. Mahon and McGowan note that Porter's model fails to recognize the importance of nonmarket factors and their impact on business strategies. See also J. Moore, *The Death of Competition,* and J. Moore, "Predators and Prey: A New Ecology of Competition," *Harvard Business Review* 71, no. 3 (1993): 75–86 for the development of the biological metaphor as a basis for competitive analysis. Moore coined the term "business ecosystem" to describe the coevolving community of firms that both cooperate and compete. Given the name he used for his model, it is interesting that Moore ignores the natural environment in his analysis.

16. See P. Ehrlich and P. Raven, "Butterflies and Plants: A Study in Coevolution," *Evolution* 18 (1964): 586–608.

17. See D. Futuyma and M. Slatkin, *Coevolution* (Sunderland, MA: Sinauer Associates, 1983).

18. See Moore, *The Death of Competition*; and Moore, "Predators and Prey."

19. See T. Rowley, "Moving Beyond Dyadic Ties: A Network Theory of Stakeholder Influences," *Academy of Management Review* 22, no. 4 (1997): 887–910.

20. See R. Fried, "The State of Environment and Business," *Sustainable Business*, June 1999; available at sustainablebusiness.com/html/insider/june99/state.cfm.

21. See M. Starik and G. Rands, "Weaving an Integrated Web: Multilevel and Multisystem Perspectives of Ecologically Sustainable Organizations," *Academy of Management Review* 20, no. 4 (1995): 908–35.

22. For a review of the process of scenario development at Royal/Dutch Shell, see P. Swartz, *The Art of the Long View* (New York: Doubleday, 1991); P. Wack, "Scenarios: Uncharted Waters," *Harvard Business Review* 63, no. 5 (1985): 73–89; and P. Wack, "Scenarios: Shooting the Rapids," *Harvard Business Review* 63, no. 6 (1985): 139–50.

23. Our arguments in this section are taken largely from D. Bunn and A. Salo, "Forecasting with Scenarios," *European Journal of Operational Research* 68 (1993): 291–303; and S. Millet, "How Scenarios Trigger Strategic Thinking," *Long Range Planning* 21, no. 5 (1988): 61–68.

24. For an excellent discussion of the institutional learning resulting from scenario building, see M. Tenaglia and P. Noonan, "Scenario-Based Strategic Planning: A Process for Building Top Management Consensus," *Planning Review* 20, no. 2 (1992): 8–12.

Notes to Chapter 5

1. This section is constructed largely from the following references: J. Barney, "Looking Inside for Competitive Advantage," *Academy of Management Executive* 9 (1995): 49–61; J. Duncan, P. Ginter, and L. Swayne, "Competitive Advantage and Internal Organizational Assessment," *Academy of Management Executive* 12 (1998): 6–16; K. Eisenhardt and J. Martin, "Dynamic Capabilities: What Are They?" *Strategic Management Journal* 21 (2000):1105–21; E. Freeman, *Strategic Management: A Stakeholder Approach* (Boston: Pitman, 1984); C. Prahalad and G. Hamel, "The Core Competence of the Organization," *Harvard Business Review* (May–June 1990): 79–91; J. Pearce and R. Robinson, *Strategic Management* (New York: McGraw-Hill, 2003).

2. For a thorough discussion of the application of the resource-based view to sustainable strategic management, see S. Hart, "A Natural Resource-Based View of the Firm," *Academy of Management Review* 20, no. 4 (1995): 966–1014; S. Hart, "Beyond Greening: Strategies for a Sustainable World," *Harvard Business Review* (January–February 1997): 67–76; and M. Russo and P. Fouts, "A Resource-Based Perspective on Corporate Environmental Performance and Profitability," *Academy of Management Journal* 40, no. 3 (1997): 534–59.

3. For a thorough discussion of the value chain, see M. Porter, *Competitive Strategy* (New York: Free Press, 1980).

4. This view of the value chain is put forth by E. Freeman and J. Liedtka, "Stakeholder Capitalism and the Value Chain," *European Management Journal* 15, no. 3 (1997): 289–99.

5. See the classic work by Freeman, *Strategic Management*, for a thorough discussion of stakeholder theory.

6. See Freeman and Liedtka, "Stakeholder Capitalism and the Value Chain"; and J. Mahon and R. McGowan, "Modeling Industry Political Dynamics," *Business and Society* 37, no. 4 (1998): 391.

7. For a thorough discussion of both ecoefficiency and ecoeffectiveness presented in this and following paragraphs, see W. McDonough and M. Braungart, *Cradle to Cradle* (New York: North Point Press, 2002). (Note that their book is actually made from a technical nutrient rather than paper.)

8. For a discussion of both socioefficiency and socioeffectiveness presented in this and following paragraphs, see T. Dyllick and K. Hockerts, "Beyond the Business Case for Corporate Sustainability," *Business Strategy and the Environment* 11 (2002): 130–41; and K. Hockerts, "SustainAbility Radar," *Greener Management International* (Spring 1999): 25–35.

9. See McDonough and Braungart, *Cradle to Cradle*.

10. See Freeman and Liedtka, "Stakeholder Capitalism and the Value Chain."

11. See Dyllick and Hockerts, "Beyond the Business Case for Corporate Sustainability"; and Hockerts, "SustainAbility Radar."

12. See S. Sharma and H. Vredenburg, "Proactive Corporate Environmental Strategy and the Development of Competitively Valuable Organizational Capabilities," *Strategic Management Journal* 19 (1998): 729–53.

13. For further discussion of the earth as a stakeholder, see the following: M. Starik, "Essay: The Toronto Conference: Reflections on Stakeholder Theory," *Business and Society* 33, no. 1 (1994): 89–95; M. Starik, "Should Trees Have Managerial Standing? Toward Stakeholder Status for Non-Human Nature," *Journal of Business Ethics* 14 (1995): 207–17; E. Stead and J. Stead, "Earth: A Spiritual Stakeholder," in *Environmental Challenges to Business*, ed. J. Reichart and P. Werhane, 231–44 (Bowling Green, OH: Society for Business Ethics, 2000); and J. Stead and E. Stead, "Eco-Enterprise Strategy: Standing for Sustainability," *Journal of Business Ethics* 24, no. 4 (2000): 313–29.

14. For a thorough discussion of stakeholder theory, see Freeman, *Strategic Management*. The definition quoted herein can be found on page 46 of that classic work.

15. See I. Henriques and P. Sadorsky, "The Relationship Between Environmental Commitment and Managerial Perceptions of Stakeholder Importance," *Academy of Management Journal* 42, no. 1 (1999): 87–99.

16. The discussion of social legislation in this and following paragraphs in this section is based on information found in A. Carroll and A. Buchholtz, *Business and Society* (Mason, OH: South-Western, 2003). For a more complete discussion of the evolution of the importance of nonmarket critical success factors in strategic management, please see I. Ansoff, "The Changing Shape of the Strategic Problem," in *Strategic Management*, ed. D. Schendel and C. Hofer, 30–44 (Boston: Little Brown, 1979).

17. This information is from *Business and Human Rights* (London: Royal Dutch Shell, 1999), which contains an excellent in-depth discussion of human rights and the role of business.

18. For a detailed discussion of these cases, see W. Stead, M. McKinney, and J. Stead, "Institutionalizing Environmental Performance in U.S. Industry: Is It Happen-

ing and What If It Does Not?" *Business Strategy and the Environment* 7, no. 5 (1998): 261–70.

19. See M. Porter and C. van der Linde, "Green and Competitive: Ending the Stalemate," *Harvard Business Review* 73, no. 5 (1995): 120–34.

20. For an excellent discussion of Project XL and its effectiveness, see A. Marcus, D. Geffen, and K. Sexton, *Reinventing Environmental Regulation: Lessons from Project XL* (Washington, DC: RFF Press, 2002).

21. These three consumer categories were developed by: J. Ottman, *Green Marketing* (Lincolnwood, IL: NTC Contemporary, 1998). Other informative books about socially and environmentally responsible consumers include: W. Coddington, *Environmental Marketing* (New York: McGraw-Hill, 1993); and D. Fuller, *Sustainable Marketing* (Thousand Oaks, CA: Corwin Press, 1999).

22. Our discussion is based on the following: J. Makower, "Post-Mortem for Green Consumerism," *Business Ethics* 9, no. 4 (1995): 52; and Ottman, *Green Marketing*.

23. See Carroll and Buchholtz, *Business and Society*.

24. The history of ethical investment presented here is from: Council on Economic Priorities, *The Better World Investment Guide* (New York: Prentice Hall, 1991).

25. The data reported regarding the two types of ethical investing funds are from: J. Brill and A. Reder, *Investing from the Heart* (New York: Crown, 1992); "Social Investment Forum's 1999 Trends Report," *Green Money Journal* (Summer 2000); available at www.greenmoneyjournal.com/article.mpl?newsletterid=18&articleid=149; and "What Is Ethical Investment" (2001): www.web.net/ethmoney/what.htm [site no longer available].

26. The information on CERES is from its Web site: www.ceres.org. The site contains excellent information about the work of this organization. We strongly encourage readers to take time to visit this site and to learn more about CERES, its partners, and its endorsers.

27. This discussion is based on Freeman, *Strategic Management*.

28. S. Hoe, *The Man Who Gave His Company Away* (London: William Heinemann, 1978), 80. This book provides a thorough discussion of the life and ideas of Ernest Bader.

29. The information regarding E.F. Schumacher's concerns for employees is from: Hoe, *The Man Who Gave His Company Away*. The Schumacher quote included in this discussion is from E.F. Schumacher, *Good Work* (New York: Harper and Row, 1979), 119; E.F. Schumacher, *Small Is Beautiful: Economics as if People Mattered* (New York: Harper and Row, 1973); and B. Wood, *E.F. Schumacher: His Life and Thought* (New York: Harper and Row, 1984).

30. For an excellent discussion of employee rights, see Carroll and Buchholtz, *Business and Society*.

31. For a thorough discussion of the role of human capital in sustainable strategic management, see D. Dunphy, A. Griffiths, and S. Benn, *Organizational Change for Corporate Sustainability* (London: Routledge, 2002).

32. Many organizations now publish information on their sustainable supplier relationships in their annual social responsibility reports. Four such reports were used as references in this section: Royal Dutch Shell, *Business and Human Rights*; "Engage, Innovate, Transform," *AOL Time Warner 2002 Social Responsibility Report* (New York: AOL Time Warner, 2002); "How Do We Stand? People, Planet and Profits," *Shell Report 2000* (London: Royal Dutch Shell, 2000); and *2002 Worldwide Social Responsibility Report* (Oak Brook, IL: McDonald's Corporation, 2002). Data

from R. Fried, "The State of Environment and Business," *Progressive Investor* (April 16, 2002): www.sustainablebusiness.com/features/feature_template.cfm?ID=808 were also used in this section.

33. The information on Amnesty International and the Southern Poverty Law Center is from their respective Web sites: www.amnesty.org and www.splcenter.org.

34. The information in this paragraph came from the following sources: J. Clair, J. Milliman, and I. Mitroff, "Clash or Cooperation? Understanding Environmental Organizations and Their Relationship to Business," in *Research in Corporate Social Performance and Policy, Supplement 1*, ed. J. Post, volume eds. D. Collins and M. Starik, 163–93 (Greenwich, CT: JAI Press, 1995); M-F. Turcotte, "Conflict and Collaboration: The Interfaces Between Environmental Organizations and Business Firms," in Post, *Research in Corporate Social Performance and Policy*, 195–229.

35. Sources used in this paragraph include: D. Amoruso, "Urban Silk Purses from Sows' Ears Made Possible by Environmental Insurance," *Realty Times* (2000), realtytimes.lycos.com/renews/20000623_environment.htm; and R. Heath, "Insurers Taking to Environmental Liability Policies," *Sacramento Business Journal*, November 28, 1997; available at sacramento.bizjournals.com/sacramento/stories/1997/12/01/focus3.html.

36. For a thorough review of the sixteen principles in the ICC Business Charter for Sustainable Development, see K. North, *Environmental Business Management* (Geneva, Switzerland: International Labour Office, 1992).

37. See "ISO 14000 Update," *Business and the Environment* 7, no. 3 (March 2001): Cutter Information Group.

38. See J. Ranganathan, "Sustainability Rulers: Measuring Corporate Environmental and Social Performance," *Sustainable Enterprise Perspectives* (Washington, DC: World Resources Institute, May 1998); available at www.wri.org/wri/meb/sei/state.html.

Notes to Chapter 6

1. For an overview of the various strategy formulation processes, see C. Lindblom, "The Science of Muddling Through," *Public Administration Review* 19, no. 2 (1959): 76–88; H. Mintzberg, *The Rise and Fall of Strategic Planning* (New York: Free Press, 1994); and J. Quinn, "Logical Incrementalism," *Sloan Management Review* 30, no. 4 (1989): 45–61.

2. For an informative discussion of life-cycle analysis, see S. Svoboda, "Note on Life Cycle Analysis," in *Environmental Management*, ed. M. Russo, 217–27 (Boston: Houghton Mifflin, 1999).

3. See C. Cone, "Cause Branding in the 21st Century," *Cone and Roper Cause Related Trends Report: The Evolution of Cause Branding* (Boston: Cone, 1999); and M. Drumwright, "Company Advertising with a Social Dimension: The Role of Noneconomic Criteria," *Journal of Marketing* 60 (1996): 71–87.

4. See S. Waddock, "Stakeholder Relationships and Corporate Responsibility," paper presented at the Academy of Management Annual Meeting, Seattle, 2003. BP Amoco, Nike, McDonald's, Royal Dutch Shell, and Bristol-Myers Squibb are just a few of the firms reporting on their sustainability performance.

5. Together these references provide an excellent perspective on the development of the concept of strategic environmental management: L. Anderson and T. Bateman, "Individual Environmental Initiatives: Championing Natural Environmental Issues in Business Organizations," *Academy of Management Journal* 43, no. 4 (2000): 548–70; P. Bansal and K. Roth, "Why Companies Go Green: A Model of Ecological Responsiveness," *Academy of Management Journal* 43, no. 4 (2000): 717–36; C. Christmann, "Effects of Best Practices of Environmental Management on Cost Advantage: The Role of Complementary Assets," *Academy of Management Journal* 43, no. 4 (2000): 663–80; M. Cordano and I. Frieze, "Pollution Reduction Preferences of U.S. Environmental Managers: Applying Ajzen's Theory of Planned Behavior," *Academy of Management Journal* 43, no. 4 (2000): 627–41; C. Egri and S. Herman, "Leadership in the North American Environmental Sector: Values, Leadership Styles, and Contexts of Environmental Leaders and Their Organizations," *Academy of Management Journal* 43, no. 4 (2000): 571–604; T. Gladwin, J. Kennelly, and T. Krause, "Shifting Paradigms for Sustainable Development," *Academy of Management Review* 20, no. 4 (1995): 874–907; S. Hart, "A Natural Resource-Based View of the Firm," *Academy of Management Review* 20, no. 4 (1995): 966–1014; A. Hoffman, *Competitive Environmental Strategy* (Washington, DC: Island Press, 2000); D. Jennings and P. Zanderbergen, "Ecologically Sustainable Organizations: An Institutional Approach," *Academy of Management Review* 20, no. 4 (1995): 1015–32; A. King, "Avoiding Ecological Surprise: Lessons from Long-Standing Communities," *Academy of Management Review* 20, no. 4 (1995): 961–85; J. Maxwell, S. Rothenberg, F. Briscoe, and A. Marcus, "Green Schemes: Corporate Environmental Strategies and Their Implementation," *California Management Review* 39, no. 3 (1997): 118–32; W. McDonough and M. Braungart, *Cradle to Cradle* (New York: North Point Press, 2002); B. Piasecki, K. Fletcher, and F. Mendelson, *Environmental Management and Business Strategy* (New York: Wiley, 1995); J. Post, "Managing as if the Earth Mattered," *Business Horizons* (July/August 1991): 32–38; J. Post and B. Altman, "Models for Corporate Greening: How Corporate Social Policy and Organizational Learning Inform Leading-Edge Environmental Management," in *Research in Corporate Social Policy and Performance*, ed. J. Post, 3–29 (Greenwich, CT: JAI Press, 1992); C. Ramus and U. Steger, "The Roles of Supervisor Support Behaviors and Environmental Policy in Employee Ecoinitiatives at Leading-Edge European Companies," *Academy of Management Journal* 43, no. 4 (2000): 605–26; N. Roome, ed., *Sustainability Strategies for Industry* (Washington, DC: Island Press, 1998); S. Sharma and H. Vredenburg, "Proactive Corporate Environmental Strategy and the Development of Competitively Valuable Organizational Capabilities," *Strategic Management Journal* (1998): 729–53; P. Shrivastava, "The Role of Corporations in Achieving Ecological Sustainability," *Academy of Management Review* 20, no. 4 (1995): 936–60; M. Starik and G. Rands, "Weaving an Integrated Web: Multilevel and Multisystem Perspectives of Ecologically Sustainable Organizations," *Academy of Management Review* 20, no. 4 (1995): 908–35; W. Stead and J. Stead, "An Empirical Investigation of Sustainability Strategy Implementation in Industrial Organizations," in *Research in Corporate Performance and Policy*, Supplement 1, ed. J. Post; vol. ed. D. Collins and M. Starik, 43–66 (Greenwich, CT: JAI Press, 1995); W. Stead and J. Stead, *Management for a Small Planet: Strategic Decision Making and the Environment* (Thousand Oaks, CA: Sage, 1996); G. Throop, M. Starik, and G. Rands, "Sustainable Strategy in a Greening World: Integrating the Natural Environment into Strategic Management," *Advances in Strategic Management*, vol. 9, 63–92 (Greenwich, CT: JAI, 1993); and M. Winn, "Corporate Leadership and Policies for the Natural Environment," in *Research in Corporate Performance*, Supplement 1, 127–61.

6. See M. Porter, *Competitive Advantage* (New York: Free Press 1985).

7. See Hart, "A Natural Resource–Based View of the Firm" and S. Hart, "Beyond Greening: Strategies for a Sustainable World," *Harvard Business Review* (January–February 1997): 67–76. For further discussion of pollution prevention strategies, see *Buried Treasure: Uncovering the Business Case for Corporate Sustainability* (London: SustainAbility and the United Nations Environment Programme, 2001); Christmann, "Effects of Best Practices"; Cordano and Frieze, "Pollution Reduction Preferences"; C. Frankel, *In Earth's Company* (Gabriola Island, BC: New Society, 1998); Shrivastava, "The Role of Corporations in Achieving Ecological Sustainability"; and Stead and Stead, "An Empirical Investigation of Sustainability Strategy Implementation."

8. See C. Holliday, S. Schmidheiny, and P. Watts, *Walking the Talk* (Sheffield, UK: Greenleaf, 2002).

9. See Holliday, Schmidheiny and Watts, *Walking the Talk,* and Willard, *The Sustainability Advantage*.

10. For a discussion of Xerox's remanufacturing strategies, see H. Meyer, "The Greening of Corporate America," *Journal of Business Strategy* (January/February 2000): 38–43; and J. Nash, "Beyond Compliance: The Sustainability Advantage," *Occupational Hazards* (June 2000): 31–38.

11. For further discussion of market-driven, product stewardship strategies, see *Buried Treasure*; Hart, "A Natural Resource-Based View of the Firm"; Hart, "Beyond Greening"; F.L. Reinhardt, "Bringing the Environment Down to Earth," *Harvard Business Review* (July–August, 1999): 149–57; P. Shrivastava, *Greening Business* (Cincinnati, OH: Thompson Executive Press, 1996); and Stead and Stead, "An Empirical Investigation of Sustainability Strategy Implementation."

12. For a discussion of environmental differentiation, see *Buried Treasure*; Reinhardt, "Bringing the Environment Down to Earth"; and Willard, *The Sustainability Advantage*.

13. The information on ecolabeling is from Holliday, Schmidheiny, and Watts, *Walking the Talk*.

14. See Frankel, *In Earth's Company*.

15. The data presented in this section are from *Tomorrow's Markets* (Washington, DC: World Resources Institute, United Nations Environment Programme, World Business Council for Sustainable Development, 2002).

16. For other examples of environmental differentiation, see Holliday, Schmidheiny, and Watts, *Walking the Talk*; and Willard, *The Sustainability Advantage*.

17. For a discussion of socioefficiency, see T. Dyllick and K. Hockerts, "Beyond the Business Case for Corporate Sustainability," *Business Strategy and the Environment* 11 (2002): 130–41; and K. Hockerts, "SustainAbility Radar," *Greener Management International* (Spring 1999): 25–35.

18. See Holliday, Schmidheiny, and Watts, *Walking the Talk*, 105.

19. See Holliday, Schmidheiny, and Watts, *Walking the Talk*; and Willard, *The Sustainability Advantage*.

20. Holliday, Schmidheiny, and Watts, *Walking the Talk,* provides many examples similar to that of SC Johnson.

21. The information in this paragraph is from Holliday, Schmidheiny, and Watts, *Walking the Talk*.

22. See *Tomorrow's Markets*.

23. The information in this paragraph is from C. K. Prahalad and S. Hart, "The

Fortune at the Bottom of the Pyramid," *Strategy + Business*, First Quarter 2002; available at www.changemakers.net/library/temp/fortunepyramid.cfm.

24. See C. Murphy, "The Hunt for Globalization that Works," *Fortune*, October 28, 2002: 163–76.

25. The information in this paragraph is primarily from *Tomorrow's Markets*.

26. Unilever is just one of many examples that can be found in Prahalad and Hart, "The Fortune at the Bottom of the Pyramid"; and Willard, *The Sustainability Advantage*.

27. See Prahalad and Hart, "The Fortune at the Bottom of the Pyramid"; and Willard, *The Sustainability Advantage*.

28. See C. Frankel, "Storming the Digital Divide," *Tomorrow Magazine* 11, no. 1 (January–February 2001): 42–45; and Prahalad and Hart, "The Fortune at the Bottom of the Pyramid."

29. See *Tomorrow's Markets*.

30. See Frankel, "Storming the Digital Divide."

31. See Prahalad and Hart, "The Fortune at the Bottom of the Pyramid."

32. See *Tomorrow's Markets*.

33. See Holliday, Schmidheiny, and Watts, *Walking the Talk*.

34. P. Hawken, A. Lovins, and L. Lovins, "A Road Map for Natural Capitalism," *Harvard Business Review* (May–June 1999): 146.

35. The examples in this paragraph are from Willard, *The Sustainability Advantage*.

36. For a discussion of shared savings, see Willard, *The Sustainability Advantage*.

Notes to Chapter 7

1. For an overview of the various matrixes used in strategic decision making, see the following: F. David, *Strategic Management Concepts*, 9th ed. (Upper Saddle River, NJ: Prentice-Hall, 2003); C. Hill and G. Jones, *Strategic Management*, 5th ed. (Boston, MA: Houghton Mifflin, 2001); J. Pearce and R. Robinson, *Strategic Management*, 8th ed. (New York: McGraw-Hill, 2003); A. Thompson and A. Strickland, *Strategic Management*, 13th ed. (New York: McGraw-Hill, 2003).

2. The explanation of human cognitive processes in this section is constructed largely from the following references: M. Finney and I. Mitroff, "Strategic Plan Failures: The Organization as Its Own Worst Enemy," in *The Thinking Organization*, ed. P. Sims and D. Gioia, 317–35 (San Francisco: Jossey-Bass, 1986); D. Gioia, "Conclusion: The State of the Art in Organizational Social Cognition," in Sims and Gioia, eds., *The Thinking Organization*, 336–56; E. Rosch and B. Lloyd, *Cognition and Categorization* (Hillsdale, NJ: Lawrence Erlbaum, 1978); J. Walsh and G. Ungson, "Organizational Memory," *Academy of Management Review* 16, no. 1 (1991): 57–91; K. Weick and M. Bougon, "Organizations as Cognitive Maps," 103–33. For an excellent discussion of the role of managerial perceptions of environmental issues and strategic choice, see S. Sharma, "Managerial Interpretations and Organizational Context as Predictors of Corporate Choice of Environmental Strategy," *Academy of Management Journal* 43, no. 4 (2000): 681–97.

3. For excellent discussions related to the cognitive bases of performance appraisal, see D. Ilgen and J. Feldman, "Performance Appraisal: A Process Focus," *Research in Organizational Behavior* 5 (1983): 141–97; and J. Jolly, T. Reynolds, and J.

Slocum, "Application of the Means-End Theoretic for Understanding the Cognitive Bases of Performance Appraisal," *Organizational Behavior and Human Decision Processes* 41 (1988): 153–79.

4. The roles of values in human and organizational behavior have been covered in many excellent references. Some of those used here include: J. Liedtka, "Value Congruence: The Interplay of Individual and Organizational Value Systems," *Journal of Business Ethics* 8 (1989): 805–15; and M. Rokeach, *Beliefs, Attitudes, and Values* (San Francisco: Jossey-Bass, 1968).

5. For further discussion of social values, see E. Ravlin and B. Meglino, "Effect of Values on Perception and Decision Making: A Study of Alternative Work Values Measures," *Journal of Applied Psychology* 72, no. 4 (1987): 666–73.

6. This paragraph is based on information contained in the following references: A. Denisi, T. Cafferty, and B. Meglino, "A Cognitive View of the Performance Appraisal Process: A Model and Research Proposition," *Organizational Behavior and Human Performance* 33 (1984): 360–96; J. Gutman, "A Means-End Chain Model Based on Consumer Categorization Processes," *Journal of Marketing* 46 (1982): 60–72; and M. Rosenberg, "Cognitive Structure and Attitudinal Effect," *Journal of Abnormal and Social Psychology* 53 (1956): 367–72.

7. Several references include thorough discussions of the role of values in strategic decisions. Among them are: I. Ansoff, *Corporate Strategy: An Analytical Approach to Business Policy for Growth and Expansion* (New York: McGraw-Hill, 1980); C. Schwenk, "Cognitive Simplification Processes in Strategic Decision-Making," *Strategic Management Journal* 5 (1984): 111–28; and C. Schwenk, *The Essence of Strategic Decision Making* (Lexington, MA: Lexington Books, 1988). For an excellent discussion of environmental, ethical decision making, see B. Flannery and D. May, "Environmental Ethical Decision Making in the U.S. Metal-Finishing Industry," *Academy of Management Journal* 43, no. 4 (2000): 642–62.

8. This section is based on ideas expressed in R. Ornstein and P. Ehrlich, *New World, New Mind* (New York: Touchstone, 1989).

9. The arguments in this section are based primarily on the following sources: H. Daly, *Steady-State Economics*, 2d ed. (Washington, DC: Island Press, 1991); P. Drucker, *The New Realities* (New York: Harper and Row, 1989); A. Etzioni, *The Moral Dimension: Toward a New Economics* (New York: Free Press, 1988); and E.F. Schumacher, *Small Is Beautiful: Economics as if People Mattered* (New York: Harper and Row, 1973).

10. For further discussion of these values as they relate to sustainability, see J. Stead and E. Stead, "Eco-Enterprise Strategy: Standing for Sustainability," *Journal of Business Ethics* 24, no. 4 (2000): 313–29.

11. Ideas in this section reflect some excellent sources related to posterity. We recommend the following for those interested in further pursuit of this topic: T. Gladwin, J. Kennelly, and T. Krause, "Shifting Paradigms for Sustainable Development," *Academy of Management Review* 20, no. 4 (1995): 874–907; L. Milbrath, *Envisioning a Sustainable Society* (Albany: State University of New York Press, 1989); and Ornstein and Ehrlich, *New World, New Mind*.

12. Many of the ideas presented in this section are based on the following: Etzioni, *The Moral Dimension*; G. Hardin, "The Tragedy of the Commons," *Science* 162 (1968): 1243–48; A. King, "Avoiding Ecological Surprise: Lessons from Long-Standing Communities," *Academy of Management Review* 20, no. 4 (1995): 961–85; M. Starik and G. Rands, "Weaving an Integrated Web: Multilevel and Multisystem Perspectives of

Ecologically Sustainable Organizations," *Academy of Management Review* 20, no. 4 (1995): 908–35.

13. This discussion is based largely on the following: H. Daly and J. Cobb, Jr., *For the Common Good*, 2d ed. (Boston: Beacon Press, 1994); and Schumacher, *Small Is Beautiful*.

14. For excellent discussions of diversity from biological, social, and economic perspectives, see W. Frederick, *Values, Nature, and Culture in the American Corporation* (New York: Oxford University Press, 1995); J. Lovelock, *The Ages of Gaia* (New York: Bantam Books, 1988); J.F. Moore, *The Death of Competition: Leadership and Strategy in the Age of Business Ecosystems* (New York: HarperBusiness, 1996); G. Robinson and K. Dechant, "Building a Business Case for Diversity," *Academy of Management Executive* 11 (1997): 21–31; and E.O. Wilson, *The Diversity of Life* (New York: Norton, 1992).

15. See Daly, *Steady-State Economics*; and E.F. Schumacher, *A Guide for the Perplexed* (New York: Harper and Row, 1977).

16. Prominent scholars who have focused their attention on the need for spiritual fulfillment in the workplace include: S. Covey, *The 7 Habits of Highly Effective People* (New York: Simon and Schuster, 1990); W. Halal, *The New Capitalism* (New York: Wiley, 1986); C. Handy, *The Age of Unreason* (Boston: Harvard Business School Press, 1989); A. Maslow, *Toward a Psychology of Being* (Princeton, NJ: Van Nostrand Reinhold, 1962); Schumacher, *Small Is Beautiful*; and P. Senge, *The Fifth Discipline: The Art and Practice of the Learning Organization* (New York: Doubleday/Currency, 1990).

17. This model is a broad interpretation of Stead's and Stead eco-enterprise strategy model, which can be found in Stead and Stead, "Eco-Enterprise Strategy."

Notes to Chapter 8

1. For further discussions of strategy implementation and evaluation, please see the following: F. David, *Strategic Management Concepts*, 9th ed. (Upper Saddle River, NJ: Prentice-Hall, 2003); C. Hill and G. Jones, *Strategic Management*, 5th ed. (Boston: Houghton Mifflin, 2001); J. Pearce and R. Robinson, *Strategic Management*, 8th ed. (New York: McGraw-Hill, 2003); A. Thompson and A. Strickland, *Strategic Management*, 13th ed. (New York: McGraw-Hill, 2003).

2. For more information on the Natural Step, see B. Nattrass and M. Altomare, *The Natural Step for Business: Wealth, Ecology and the Evolutionary Corporation* (Gabriola Island, BC: New Society, 2002).

3. For a detailed discussion of the lessons that businesses can learn from nature, see T. Kiuchi and B. Shireman, *What We Learned in the Rainforest* (San Francisco: Barrett-Koehler Publishers, 2002).

4. For further discussions of life-cycle assessment, see the following references: D. Ciambrone, *Environmental Life Cycle Analysis* (Boca Raton: Lewis, 1997); T. Gladwin, S. Levin, and J. Ehrenfeld, "Research, Development and Industrial Testing of a Sustainability Impact Assessment System," a proposal made in cooperation with AT&T, Bristol-Myers Squibb, Digital Equipment Corporation, Merck & Co., and Philips Electronics to the National Science Foundation, 1994; S. Svoboda, "Note on Life Cycle Analysis," in *Environmental Management*, ed. M. Russo, 217–27 (Boston: Houghton Mifflin, 1999); and D. Thompson, *Tools for Environmental Management* (Gabriola Island, BC: New Society, 2002).

5. See H. Meffert and M. Kirchgeorg, "Market-oriented Environmental Management: Challenges and Opportunities for Green Marketing," Working Paper no. 43 (Munster, Germany: Institut fur Marketing der Westfalischen Wilhelms-Universitat Munster, 1995); and M. Sullivan and J. Ehrenfeld, "Reducing Life-cycle Environmental Impacts: An Industry Survey of Emerging Tools and Programs," in *Environmental TQM*, ed. J. Willig, 43–49 (New York: McGraw-Hill, 1994).

6. See P. Dillon, "Implications of Industrial Ecology in Firms," in *The Greening of Industrial Ecosystems*, ed. R. Allenby and D. Richards, 201–7 (Washington, DC: National Academy Press, 1994).

7. See B. Allenby, "Integrating Environment and Technology: Design for Environment," in *The Greening of Industrial Ecosystems*, 147–48; and D. Richards, B. Allenby, and R. Frosch, "The Greening of Industrial Ecosystems: Overview and Perspective," in *The Greening of Industrial Ecosystems*, 1–19.

8. The information in this paragraph is from Allenby, "Integrating Environment and Technology"; and Gladwin, Levin, and Ehrenfeld, "Research, Development and Industrial Testing of a Sustainability Impact Assessment System."

9. See A. King, "Improved Manufacturing Resulting from Learning-from-Waste: Causes, Importance and Enabling Conditions," paper presented at the Academy of Management Annual Meeting, Dallas, 1994.

10. See Allenby, "Integrating Environment and Technology"; and Richards, Allenby, and Frosch, "The Greening of Industrial Ecosystems: Overview and Perspective."

11. For a discussion of technical and biological nutrients and their role in design, see W. McDonough and M. Braungart, *Cradle to Cradle* (New York: North Point Press, 2002).

12. See *Design for Environment: A Case Study of the Power Mac G4 Desktop Computer* (Cupertino, CA: Apple Environmental Technologies and Strategies, Apple Computer, March 2000); and T. Oyama and E. Hirose, "Approaches to Design for the Environment with Practical Examples," *Mitsubishi Electric ADVANCE* (December 2001): 25–27.

13. See J. Fiksel, "Emergence of a Sustainable Business Community," *Pure and Applied Chemistry* 73, no. 8 (2001): 1265–68.

14. For a detailed discussion of the lessons that businesses can learn from nature, see Kiuchi and Shireman, *What We Learned in the Rainforest*. For more information on the Natural Step, see Nattrass and Altomare, *The Natural Step for Business*.

15. See J. Greeno, "Corporate Environmental Excellence and Stewardship," in *Environmental Strategies Handbook*, ed. R. Kolluru, 43–64 (New York: McGraw-Hill, 1994); and P. Shrivastava, *Greening Business* (Cincinnati, OH: Thompson Executive Press, 1996).

16. See B. Willard, *The Sustainability Advantage* (Gabriola Island, BC: New Society, 2002).

17. For further information on issues in sustainable procurement, see Thompson, *Tools for Environmental Management*.

18. See J. Makover, ed., *The Green Business Letter* (Oakland, CA: Tilden Press, August 2001).

19. For an excellent discussion on educating employees about sustainable consumption of material and energy flows within the organization, see Nattrass and Altomare, *The Natural Step for Business*.

20. See P. Hawken, A. Lovins, and L. Lovins, "A Road Map for Natural Capitalism," *Harvard Business Review* (May–June 1999): 145–57.

21. Data were obtained Willard, *The Sustainability Advantage*.

22. For examples of other firms utilizing these innovative strategies, see Hawken, Lovins, and Lovins, "A Road Map for Natural Capitalism"; C. Holliday, S. Schmidheiny, and P. Watts, *Walking the Talk* (Sheffield, UK: Greenleaf, 2002); and Willard, *The Sustainability Advantage*.

23. Research by R. Klassen and B. Whybark, "The Impact of Environmental Technologies on Manufacturing Performance," *Academy of Management Journal* 42, no. 6 (1999): 599–615, empirically demonstrates the effectiveness of ecoefficiency technologies. Perhaps no other book on the topic of ecoefficiency has been as influential in the past decade as has P. Hawken, A. Lovins, and L. Lovins, *Natural Capitalism: Creating the Next Industrial Revolution* (Boston: Little, Brown, 1999). The latter two coauthors are founders of the Rocky Mountain Institute, which specializes in these and related topics, so readers are referred to their regularly updated Web site for more information: www.rmi.org. For further information on ecoefficiency, see the following: Holliday, Schmidheiny, and Watts, *Walking the Talk*; McDonough and Braungart, *Cradle to Cradle*; and Willard, *The Sustainability Advantage*.

24. Pollution prevention became a very common environmental management strategy in the 1990s, with many businesses, governments, and nonprofit organizations participating in its development and practice worldwide. For further discussion of pollution prevention strategies, see *Buried Treasure: Uncovering the Business Case for Corporate Sustainability* (London: SustainAbility and the United Nations Environment Programme, 2001); C. Christmann, "Effects of Best Practices of Environmental Management on Cost Advantage: The Role of Complementary Assets," *Academy of Management Journal* 43, no. 4 (2000): 663–80; M. Cordano and I. Frieze, "Pollution Reduction Preferences of U.S. Environmental Managers: Applying Ajzen's Theory of Planned Behavior," *Academy of Management Journal* 43, no. 4 (2000) 627–41; C. Frankel, *In Earth's Company* (Gabriola Island, BC: New Society, 1998); S. Hart, "A Natural Resource-Based View of the Firm," *Academy of Management Review* 20, no. 4 (1995): 966–1014; "Beyond Greening: Strategies for a Sustainable World," *Harvard Business Review* (January–February 1997): 67–76; J. Romm, *Lean and Clean Management* (New York: Kodansha, 1994); P. Shrivastava, "The Role of Corporations in Achieving Ecological Sustainability," *Academy of Management Review* 20, no. 4 (1995): 936–60; and W. Stead and J. Stead, "An Empirical Investigation of Sustainability Strategy Implementation in Industrial Organizations," in *Research in Corporate Performance and Policy, Supplement 1*, ed. J. Post, vol. ed. D. Collins and M. Starik, 43–66 (Greenwich, CT: JAI, 1995).

25. For more examples of firms utilizing pollution prevention strategies, see Holliday, Schmidheiny, and Watts, *Walking the Talk*; and Willard, *The Sustainability Advantage*.

26. For more information on energy efficiency and energy substitution, see Holliday, Schmidheiny, and Watts, *Walking the Talk*; and Willard, *The Sustainability Advantage*.

27. For further information on TQEM, see N. Ahmed, "Incorporating Environmental Concerns into TQM," *Production and Inventory Management Journal* (First Quarter 2001): 25–30; J. Nash, K. Nutt, J. Maxwell, and J. Ehrenfeld, "Polaroid's Environmental Accounting and Reporting System: Benefits and Limitations of a TQEM

Measurement Tool," in *Environmental TQM*, ed. J. Willig, 217–34 (New York: McGraw-Hill, 1994); and K. North, *Environmental Business Management* (Geneva: International Labour Organization, 1997).

28. For an excellent discussion of how ecoefficiency demonstrates a failure of the imagination and the importance of ecoeffectiveness in sustainability, see McDonough and Braungart, *Cradle to Cradle*.

29. Industrial ecology is an environmental management topic that has grown significantly throughout the 1990s. In addition to conferences and journals dedicated to this topic, an increasing number of industrial communities are experimenting with this approach. One of the most often cited sources on this topic is T.E. Graedel and B.R. Allenby, *Industrial Ecology* (Englewood Cliffs, NJ: Prentice-Hall, 2002).

30. For further information, see D. Crawford, "Sustainability Is Architect's Mantra," *Business First–Columbus*, December 8, 2000; McDonough and Braungart, *Cradle to Cradle*; D. Wann, *Deep Design: Pathways to a Livable Future* (Washington, DC: Island Press, 1999); and M. Weaver, "Grimshaw Sets Green Standard," *Building Design*, January 12, 2001.

31. See D. Fuller, *Sustainable Marketing* (Thousand Oaks, CA: Sage, 1999), 4.

32. For more information on green marketing, see Jacquelyn Ottman's excellent Web site: www.greenmarketing.com, and her book, *Green Marketing: Opportunity for Innovation* (New York: NTC-McGraw-Hill, 1998). See also M. Charter and M. Polonsky, eds., *Greener Marketing: A Global Perspective on Greening Marketing Practice* (Sheffield, UK: Greenleaf, 1999); and W. Coddington, *Environmental Marketing* (New York: McGraw-Hill, 1993).

33. See Meffert and Kirchgeorg, "Market-oriented Environmental Management."

34. For further information on the challenges of environmental marketing, see Ottman, *Green Marketing*.

35. See Coddington, *Environmental Marketing*; and Ottman, *Green Marketing*.

36. In *Green Marketing*, Ottman provides numerous examples of how marketers are rethinking ways to deliver value to customers via selling services.

37. See Ottman, *Green Marketing*.

38. The information on the components of a successful sustainable marketing strategy is from the following sources: J. Frankel and W. Coddington, "Environmental Marketing," in *Environmental Strategies Handbook*, ed. R.V. Kolluru, 643–77 (New York: McGraw-Hill, 1994); Fuller, *Sustainable Marketing*; and Ottman, *Green Marketing*.

39. See C. Cone, "Cause Branding in the 21st Century," *Cone and Roper Cause Related Trends Report: The Evolution of Cause Branding* (Boston: Cone, Inc., 1999); and M. Drumwright, "Company Advertising with a Social Dimension: The Role of Noneconomic Criteria," *Journal of Marketing* 60 (1996): 71–87.

40. J. Kuhre, *ISO 14020s: Efficient and Accurate Environmental Marketing Procedures* (Upper Saddle River, NJ: Prentice-Hall EGS Professional, 2002).

41. J. Ottman and M. Polonsky, "Developing Green Products: Learning from Stakeholders," *Journal of Sustainable Product Design*, April 5, 1998.

42. For a detailed discussion of sustainable pricing strategies, see Fuller, *Sustainable Marketing*.

43. For an excellent discussion of sustainable channel networks, see Fuller, *Sustainable Marketing*.

Notes to Chapter 9

1. In addition to the references related to creating sustainability-centered organizational cultures that we cited in Chapters 1 and 3, readers may also want to refer to B. Daily and S. Huang, "Achieving Sustainability Through Attention to Human Resource Factors in Environmental Management," *International Journal of Operations and Production Management* 21, no. 12 (2001): 1539–52.

2. For more detailed discussions of the role of and development of stories, myths, and legends in creating sustainability-centered cultures, see K. Starkey and A. Crane, "Toward Green Narrative: Management and the Evolutionary Epic," *Academy of Management Review* 28, no. 2 (2003): 220–37; and E. Stead and J. Stead, "Earth: A Spiritual Stakeholder," in *Environmental Challenges to Business*, ed. J. Reichart and P. Werhane, 231–44 (Bowling Green, OH: Society for Business Ethics, 2000).

3. For more information on Patagonia, see M. Winn, "Corporate Leadership and Policies for the Natural Environment," in *Research in Corporate Performance and Policy, Supplement 1*, ed. J. Post, vol. ed. D. Collins and M. Starik, 127–61 (Greenwich, CT: JAI, 1995).

4. The need for organizations to shed their machine-like structures in favor of more organic structures as environmental turbulence increases and technology advances was fostered in the following classic works: T. Burns and G. Stalker, *The Management of Innovation* (London: Tavistock, 1961); F. Emery and E. Trist, *Towards a Social Ecology: Contextual Appreciations of the Future in the Present* (New York: Plenum, 1973); J. Woodward, *Industrial Organization Theory and Practice* (Oxford, UK: Oxford University Press, 1965).

5. Schumacher's theory of large-scale organization is discussed in depth in E.F. Schumacher, *Small Is Beautiful: Economics as if People Mattered* (New York: Harper and Row, 1973). Another reference important to this discussion is E.F. Schumacher, *Good Work* (New York: Harper and Row, 1979). The direct quotes in this discussion are from *Small Is Beautiful*, 242, and *Good Work*, 83.

6. The following references are related to the need for learning structures in sustainable strategic management: W. Halal, *The New Capitalism* (New York: Wiley, 1986); J. Post and B. Altman, "Models for Corporate Greening: How Corporate Social Policy and Organizational Learning Inform Leading-Edge Environmental Management," in *Research in Corporate Social Policy and Performance*, ed. J. Post, 3–29 (Greenwich, CT: JAI Press, 1992); P. Shrivastava, "Ecocentric Management in Industrial Ecosystems: Management Paradigm for a Risk Society," *Academy of Management Review* 20, no. 1 (1995): 118–37; and W. Stead and J. Stead, "Can Humankind Change the Economic Myth? Paradigm Shifts Necessary for Ecologically Sustainable Business," *Journal of Organizational Change Management* 7, no. 4 (1994): 15–31.

7. The discussion of learning organizations is drawn primarily from the following sources: R. Beckhard and W. Pritchard, *Changing the Essence* (San Francisco: Jossey-Bass, 1992); C. Handy, *The Age of Unreason* (Boston: Harvard Business School Press, 1989); and P. Senge, *The Fifth Discipline: The Art and Practice of the Learning Organization* (New York: Doubleday/Currency, 1990).

8. These six human resource management factors were developed from three sources, each with an excellent discussion of the various human resource dimensions of sustainable strategic management: Daily and Huang, "Achieving Sustainability

through Attention to Human Resource Factors"; C. Egri and R. Hornal, "Strategic Environmental Human Resources Management and Organizational Performance: An Exploratory Study of the Canadian Manufacturing Sector," in *Research in Corporate Sustainability: The Evolving Theory and Practice of Organizations in the Natural Environment*, ed. S. Sharma and M. Starik, 205–36 (Northhampton, MA: Elgar, 2002); and B. Willard, *The Sustainability Advantage* (Gabriola Island, BC: New Society, 2002).

9. The research referred to in this paragraph is reported in Egri and Hornal, "Strategic Environmental Human Resources Management"; and Willard, *The Sustainability Advantage*.

10. These data are from Egri and Hornal, "Strategic Environmental Human Resources Management."

11. See Daily and Huang, "Achieving Sustainability Through Attention to Human Resource Factors."

12. See Egri and Hornal, "Strategic Environmental Human Resources Management"; and W. Stead and J. Stead, "An Empirical Investigation of Sustainability Strategy Implementation in Industrial Organizations," in *Research in Corporate Performance and Policy*, 43–66.

13. See Daily and Huang, "Achieving Sustainability Through Attention to Human Resource Factors"; and Egri and Hornal, "Strategic Environmental Human Resources Management."

14. Our discussion of disruptive technologies was developed on the basis of two excellent sources that expand on the idea in much more depth than we do here: S. Hart and M. Milstein, "Global Sustainability and the Creative Destruction of Industries," *Sloan Management Review* (Fall 1999): 23–33; and C. Holliday, S. Schmidheiny, and P. Watts, *Walking the Talk* (Sheffield, UK: Greenleaf, 2002).

15. See K. Adamson and T. Foxon, "Disruptive Technologies and Sustainable Development: The Case of Fuel Cell Vehicles," paper presented at the third Policies for Sustainable Technological Innovation (POSTI) International Conference, London, December 1–3, 2000; T. Peterson, "Solar Power: Updating the Market Perspective for a Disruptive Technology," *Electric Light and Power* (June 2000).

16. The difference between developing and developed countries regarding sustainable technologies is fast becoming a core sustainable development concern. Recent World Bank/International Monetary Fund protests and continuing post–September 11, 2001, frictions between some developed and developing nations keep these differences at center stage. One of the more enlightening sources of information on this topic is A. Hammond, *Which World?* (New York: Norton, 1998).

Notes to Chapter 10

1. For an overview of strategy evaluation processes, see F. David, *Strategic Management Concepts*, 9th ed. (Upper Saddle River, NJ: Prentice-Hall, 2003); C. Hill and G. Jones, *Strategic Management*, 5th ed. (Boston: Houghton Mifflin, 2001); J. Pearce and R. Robinson, *Strategic Management*, 8th ed. (New York: McGraw-Hill, 2003); A. Thompson and A. Strickland, *Strategic Management*, 13th ed. (New York: McGraw-Hill, 2003).

2. M. Arnold and R. Day, *The Next Bottom Line: Making Sustainable Development Tangible* (Washington, DC: World Resources Institute, 1998).

3. See the Delphi Group, *Environmental Performance and Competitive Advantage: A Business Guide*, Publication 3648E (Toronto: Queens Printer of Ontario, 1999), for a complete list of the indicators used in EPIs. Also see B. Willard, *The Sustainability Advantage* (Gabriola Island, BC: New Society, 2002) for a discussion of the use of EPIs in measurement and reporting.

4. Arnold and Day, *The Next Bottom Line*; M. Bennett and P. James, *The Green Bottom Line: Environmental Accounting for Management* (Sheffield, UK: Greenleaf, 1998).

5. See D. Lober, "Evaluating the Environmental Performance of Corporations," *Journal of Managerial Issues* 8, no. 2 (1996): 184–205.

6. This continuum for self-assessing sustainability performance can be found in Willard, *The Sustainability Advantage*.

7. See C. Holliday, S. Schmidheiny, and P. Watts, *Walking the Talk* (Sheffield, UK: Greenleaf, 2002); and J. Ranganathan, "Sustainability Rulers: Measuring Corporate Environmental and Social Performance," *Sustainable Enterprise Perspectives* (Washington, DC: World Resources Institute, May 1998); available at www.wri.org/wri/meb/sei/state.html.

8. The sustainability-based performance measurement and evaluation organizations and programs discussed in this paragraph and the next can be found in Ranganathan, "Sustainability Rulers."

9. See M. Mathews and J. Lockhart, "The Use of an Environmental Equity Account to Internalize Externalities"(Birmingham, UK), research paper published by the Aston Business School Research Institute, Aston University, February 2001; and S. Schaltegger, *Corporate Environmental Accounting* (Chichester, UK: Wiley, 1996).

10. See R. Roussey, "Auditing Environmental Liabilities," *Auditing: A Journal of Practice and Theory* 11, no. 1 (1992): 47–57; and J. Surma and D. Petracca, "Accounting for Environmental Costs: What's Happening in Practice," *Journal of Environmental Leadership* (First Quarter 1993): 143–52.

11. See J. Bebbington, R. Gray, C. Hibbitt, and E. Kirk, *Full Cost Accounting: An Agenda for Action* (London: Association of Chartered Certified Accountants, 2001).

12. These factors are developed in depth in W. Sherman, D. Steingard, and D. Fitzgibbons, "Sustainable Stakeholder Accounting: Beyond Complementarity and Towards Integration in Environmental Accounting," in *Research in Corporate Sustainability: The Evolving Theory and Practice of Organizations in the Natural Environment*, ed. S. Sharma and M. Starik, 257–94 (Northampton, MA: Edward Elgar, 2002).

13. For a discussion of full-cost accounting, see J. Beaumont, L. Pedersen, and B. Whitaker, *Managing the Environment* (Oxford, UK: Butterworth Heinemann, 1993); P. Elkins, M. Hillman, and R. Hutchinson, *Wealth Beyond Measure* (London: Gaia Press, 1992), from which the automobile example came; and R. Todd, "Zero-Loss Accounting Systems," in *The Greening of Industrial Ecosystems*, ed. B. Allenby and D. Richards, 191–200 (Washington, DC: National Academy Press, 1994).

14. The information in this and the following paragraphs regarding the research and recommendations of ACCA is from Bebbington et al., *Full Cost Accounting* (a 174-page report). A summary of this report can be found at the ACCA's Web site at: www.acca.co.uk. See J. Bebbington, R. Gray, C. Hibbitt, and E. Kirk, "Full Cost Accounting Principles and Practices," *Accounting and Business*, January 1, 2002; available at www.acca.co.uk/publications/accountingandbusiness/281399.

15. In addition to the citations above, ACCA has published a 78-page guide for

implementing ecological-footprint analysis: N. Chambers and K. Lewis, *Ecological Footprint Analysis: Towards a Sustainability Indicator for Business* (London: Association of Chartered Certified Accountants, 2001).

16. For a summary of the New Zealand requirements (with software download), see www.mfe.govt.nz/publications/waste/landfill-full-cost-accounting-guide-mar02.html. For Florida's full cost accounting requirements (and software), see www.dep.state.fl.us/waste/categories/fca/default.html.

17. See J. Elkington, *Cannibals with Forks* (Oxford, UK: Capstone, 1997); and R. Gale and P. Stokoe, "Environmental Cost Accounting and Business Strategy," in *Handbook of Environmentally Conscious Manufacturing*, ed. C. Madu, 119–37 (Amsterdam: Kluwer Academic, 2001).

18. For further discussion of Innovest's EcoValue'21 paltform, see Holliday, Schmidheiny, and Watts, *Walking the Talk;* and Willard, *The Sustainability Advantage*.

19. For further discussion of the DSJI, see Willard, *The Sustainability Advantage*.

20. For an example of this type of software, see Greenware's Web site: greenware.ca/software/perf1.html.

21. For further information on GEMI's SD Planner, go to their Web site: www.gemi.org. As of this writing, GEMI was providing both the software and users guide at no charge.

22. See M. Moore, "A Strategic Systems Approach to Understanding Environmental Management Information Systems," *Environmental Quality Management* (Summer 2000): 65–73.

23. This information is from B. Piasecki, K. Fletcher, and F. Mendelson, *Environmental Management and Business Strategy* (New York: Wiley, 1999).

24. These factors are from ACCA's *An Introduction to Environmental Reporting* (London: Association of Chartered Certified Accountants, 2001); and from the newsletter *Business and the Environment* 7, no. 8 (1996): 2.

25. See ACCA's *An Introduction to Environmental Reporting*.

26. For detailed information on the Global Reporting Initiative, see their excellent Web site: www.globalreporting.org. Also see ACCA's *An Introduction to Environmental Reporting*.

27. See J. Ranganathan, "Sustainability Rulers."

28. See Holliday, Schmidheiny, and Watts, *Walking the Talk*.

29. For more information on Seventh Generation, see its Web site: www.seventhgeneration.com.

Notes to Chapter 11

1. The data used in this section on ASD are from a personal interview with Anthony Flaccavento (January 22, 2003), interviews with other ASD staff members, on-site visits to the Appalachian Harvest and Sustainable Woods processing facilities, various other documents provided by Flaccavento and his staff, and information on the ASD Web site, which we encourage the reader to visit: www.appsusdev.org.

2. The data used in this section on Ford Motor Company are from Ford's very informative Web site, www.ford.com; from B. Morris, "Can Ford Save Ford?" *Fortune*, November 3, 2002, available at: www.fortune.com/fortune/ceo/articles/0,15114,390071,00.html; and from World Business Council for Sustainable Develop-

ment (WBCSD), *The Sustainable Mobility Project July 2002 Progress Report* (Hertfordshire, UK: WBCSD, July 2002).

3. For an excellent discussion of a new narrative designed to support and facilitate the implementation of a more ecologically sensitive economic paradigm, see K. Starkey and A. Crane, "Toward Green Narrative: Management and the Evolutionary Epic," *Academy of Management Review* 28, no. 2 (2003): 220–37.

About the Authors

W. Edward Stead is a professor of Management at East Tennessee State University (ETSU). He earned his B.S. and M.B.A. from Auburn University and his Ph.D. in Management from Louisiana State University in Baton Rouge. Before coming to ETSU in 1982, he held faculty positions at Western Illinois University, the University of Alabama in Birmingham, and Louisiana State University. He has written extensively for more than twenty years in the fields of social issues in management and organizations and the natural environment. His book (with Jean Garner Stead), *Management for a Small Planet: Strategic Decision Making and the Environment* (Sage, 1992 and 1996), received a *Choice* Outstanding Academic Book Award in 1992, and has been used at dozens of universities in the United States, Europe, and Asia. He was a founding member of the Organizations and the Natural Environment (ONE) Interest Group in the Academy of Management. He has served as program chair, chair-elect, and chair of ONE. He has also been an active member of the Academy's Social Issues in Management (SIM) Division, and he serves on the editorial review board of *Business Strategy and the Environment.* For the past five years, he has participated in a number of community sustainability projects in the Southern Appalachian region of the United States. He currently serves on the Board of Directors of Appalachian Sustainable Development, is a member of the Northeast Tennessee Development District's Ozone Action Partnership, and serves as a community sustainability consultant for the Clinch River Chapter of the Nature Conservancy.

Jean Garner Stead is a professor of Management at East Tennessee State University (ETSU). She earned her B.S. and M.A. from Auburn University, her M.B.A. from Western Illinois University, and her Ph.D. in Business Administration from Louisiana State University in Baton Rouge. Prior to her appointment at ETSU in 1982, she served on the faculty of Western Illinois

University. She has written extensively for more than two decades in the fields of organizations and the natural environment and social issues in management. Her book (with W. Edward Stead), *Management for a Small Planet: Strategic Decision Making and the Environment* (Sage, 1992 and 1996), received a *Choice* Outstanding Academic Book Award in 1992. In addition to her research, she was a founding member of the Organizations and the Natural Environment (ONE) Interest Group in the Academy of Management and has held several positions in both ONE and the Social Issues in Management (SIM) Division of the Academy of Management. In 1995, she received East Tennessee State University's Faculty Award for outstanding teaching. For the past five years, she has been involved in several community sustainability projects in the Southern Appalachian region of the United States, and she currently serves on the Board of Directors of Appalachian Sustainable Development. She is also a member of the Northeast Tennessee Development District's Ozone Action Partnership, and she serves on the ecoloan committee for the Clinch River Chapter of the Nature Conservancy.

Mark Starik is an associate professor of Strategic Management and Public Policy at the George Washington University School of Business and Public Management in Washington, DC. He was a founding member of the Organization and Natural Environment (ONE) Interest Group in the Academy of Management, where he has served as program chair, chair-elect, and chair of ONE.

Index

F

G